Science as a Cultural Human Right

PENNSYLVANIA STUDIES IN HUMAN RIGHTS

Bert B. Lockwood, Series Editor

A complete list of books in the series
is available from the publisher.

SCIENCE AS A CULTURAL HUMAN RIGHT

Helle Porsdam

PENN

UNIVERSITY OF PENNSYLVANIA PRESS

PHILADELPHIA

Published by
University of Pennsylvania Press
Philadelphia, Pennsylvania 19104-4112
www.upenn.edu/pennpress

Printed in the United States of America on acid-free paper
10 9 8 7 6 5 4 3 2 1

Hardcover ISBN 978-1-5128-2293-9
Ebook ISBN 978-1-5128-2294-6

A catalogue record for this book is available
from the Library of Congress.

CONTENTS

ABBREVIATIONS

AAAS	American Association for the Advancement of Science
CBD	UN Convention on Biological Diversity
CESCR	Committee on Economic, Social and Cultural Rights
ECHR	European Convention on Human Rights
IASC	International Arctic Science Committee
ICCPR	International Covenant on Civil and Political Rights
ICESCR	International Covenant on Economic, Social and Cultural Rights
IP	intellectual property
IUCN	International Union for Conservation of Nature
SAR	Scholars at Risk
SDG	sustainable development goal
SESAME	Synchrotron-Light for Experimental Science and Applications in the Middle East Laboratory
TRIPS	WTO Agreement on Trade-Related Aspects of Intellectual Property Rights
UDHR	Universal Declaration of Human Rights
UNDRIP	United Nations Declaration on the Rights of Indigenous Peoples
WHO	World Health Organization
WTO	World Trade Organization

INTRODUCTION

Science as a Cultural Human Right

One of the clear messages to emerge around the world during the COVID-19 crisis is just how important it is to understand the ways that science can assist society and, just as crucially, how society can engage with and shape science. The human right to science, outlined in the 1948 Universal Declaration of Human Rights (UDHR) and repeated in the 1966 International Covenant on Economic, Social and Cultural Rights (ICESCR), is key to answering this message. Article 27(1) of the UDHR recognizes everyone's right to "share in scientific advancement and its benefits"; Article 15(1) of the ICESCR recognizes everyone's "right to enjoy the benefits of scientific progress and its applications."[1] This right—the right to science—also requires states parties to develop and disseminate science, to respect the freedom of scientific research, and to recognize the benefits of international contacts and cooperation in the scientific field.

The right to science has never been more important. Even before the COVID-19 health crisis, it was evident that people around the world increasingly rely on science and technology in almost every sphere of their lives: from the development of medicines and the treatment of diseases, to transport, agriculture, and the facilitation of global communication. This is reflected in the United Nations 2030 agenda with its seventeen sustainable development goals that emphasize the importance of science and technology for sustainable development. At the same time, however, the value of science has been under attack, with some raising alarm at the emergence of "posttruth" societies. Unintended, because often unforeseen, consequences of emerging technologies are also perceived to be a serious risk and subject to "dual use."

Dual use refers to scientific research, especially in the life sciences, that can be used in a beneficent way but that also has potential to cause harm. For example, viral reverse engineering to discover why the Spanish influenza epidemic was so lethal is a type of dual-use research; on the one hand, critical information about pandemics can be gained, but on the other hand, a laboratory error or theft can cause widespread panic and consternation concerning the reasons for the underlying experiments.[2] Another example is editing of the human germline. While scientists working on such editing aim to maximize social and human welfare, they may well have started us on a slippery slope of nontherapeutic purposes, such as choosing specific traits for babies not directly related to health, like intelligence, height, or eye color (so-called designer babies).

The important role played by science and technology and the potential for dual use make it imperative to assess scientific research and its products not only on their scientific but also on their human rights merits. The added value of a human rights approach is more focus on the varied stakeholders in diverse societies. With relation to the right to science, such a focus requires "a form of affirmative action, that is, specific investments in science and technologies likely to benefit those at the bottom of the economical [sic] and social scale."[3] Legally speaking, as a human right, the right to science is not absolute. States may adopt specific measures to limit the conduct of science or the dissemination of scientific results in order to prevent harm or disrespect for other human rights.[4] Such measures are called for to protect vulnerable groups such as indigenous peoples and persons with disabilities from "the negative consequences of scientific testing or applications on, in particular, their food security, health or environment."[5]

Though it provides both scientists and members of the public with powerful legal and political tools to participate in science and to gain access to and share scientific knowledge and its applications, the right to science is surprisingly little known and underexplored. Too often restricted to academic exercises, or kept as the province of UN and human rights scholars, the implications of the right to science have yet to be fully developed, normatively as well as practically. Dedicating her third thematic report to the right to enjoy the benefits of scientific progress and its applications, the first UN special rapporteur in the field of cultural rights, Farida Shaheed, noted the urgency of the topic: "The scope, normative content and obligations of the State under this right," she writes in the first paragraph of her 2012 report, "remain underdeveloped while scientific innovations are changing human existence in ways that were inconceivable a few decades ago."[6]

Precisely because so many of the changes we are experiencing are the result of science and technology, we need to see the right to science as important in and of itself. The right to science has often been viewed as merely an ancillary right that supports others, such as the rights to health, development, and education. With the publication of General Comment No. 25 on science in April 2020, shortly after the World Health Organization declared COVID-19 a pandemic, this view is about to change.[7] The new comment provides authoritative guidance to states parties on how to implement the right to science, just as it offers instruction to the UN, human rights and scientific organizations, universities and scientists, and civil society as a whole on their rights and obligations with regard to this right.

General Comment No. 25 promotes and takes the right to science seriously as a *cultural* human right. "Cultural life is larger than science," the comment recognizes, "as it includes other aspects of human existence; it is however reasonable to include scientific activity in cultural life. Thus, the right of everyone to take part in cultural life includes the right of every person to take part in scientific progress and in decisions concerning its direction."[8] Science is a socially and culturally situated practice, and scientists and policymakers need to be more aware of the perception by the general public of science and science policies and on their effect on people's lives.

My point of departure is UDHR Article 27, but even more so ICESCR Article 15. Article 15(1) refers to the "benefits of science." This term encompasses four different things: (1) the material results of scientific research (medicines, vaccination, and technological instruments); (2) the scientific knowledge and information that derives from scientific activity; (3) the role of science in forming critical and responsible citizens who are able to participate fully in a democratic society; and (4) evidence-based decision-making processes in a democratic society.[9] The coronavirus pandemic has provided a clear illustration of how the benefits of science do not merely relate to getting a vaccination or having access to the best treatment available. They also concern access to knowledge that people need in order to participate in a responsible manner in democratic dialogue.

From a right-to-science perspective, "doing" science not only is an activity in which scientific professionals engage; it also includes collaborations between scientists and members of the public.[10] Through such collaborations, volunteers (known as citizen scientists) have helped make many important discoveries. According to Article 15(2), these discoveries must be disseminated for the benefit of all: "States parties should not only abstain from interfering in the freedom of individuals and institutions to develop science and diffuse its results. States must take positive steps for the advancement of science (development) and for the protection and dissemination of scientific knowledge and its applications (conservation and diffusion)."[11]

Article 15(3) obliges states parties to secure the "freedom indispensable for scientific research." This freedom includes the protection of researchers from unjustified interventions; the possibility to express themselves freely and openly on the ethical, human, social, or ecological implications of their research; the possibility of cooperating with other researchers, nationally as well as internationally; and the sharing of scientific data between researchers, with policymakers, and with the public wherever possible.[12]

Finally, Article 15(4) recognizes the importance of international cooperation. States should make it possible for scientific researchers to participate in the international scientific community, especially through facilitating travel in and out of their territory. Developed states should contribute to the growth of science and technology in developing countries, just as benefits should be shared with the international community. International cooperation is furthermore essential because of the risk of dual-use and because many of the most pressing problems today are global in scope and need to be solved globally.[13]

ICESCR Article 15 should, I argue, be read from the bottom up, starting with science as an international endeavor [Article 15(4)]. Without inspiration and input from fellow scientists, citizen scientists, and others around the world, individual scientists and their research groups will have none of those original ideas that lead to progress, in the short or long term. Without scientific freedom [15(3)], there will be no pursuit of ideas. There will be nothing to disseminate [15(2)]—and therefore nothing for the public to benefit from and share [15(1)].

One chapter of my monograph *The Transforming Power of Cultural Rights* (2019) concerns the right to science as one of four core cultural rights.[14] Taking that chapter as its starting point, the present book is dedicated solely to the right to science and its potential for humankind, as I see it, to defend the free and responsible practice of science—and at the same time to recognize that science and its applications are part of the cause as well as the cure. Crucially, as I argue in *The Transforming Power of Cultural Rights*, human rights, including cultural rights, constitute one of the few global ethical discourses that we have today. Immediate problems such as climate change, pandemics, resource management, and a just and sustainable economic order call for global solutions. As these heavily involve science and technology, we need a global, ethical, and cultural vocabulary and discursive tool that supports a common discussion among diverse communities that have no other common vocabularies. The right to science provides such a tool.

The Book's Structure and Chapters

The book consists of six chapters. Chapter 1 sets the scene for the rest of the book by exploring what we know about the right to science as a human right, what we do not know, and what we can do to make the right to science more known to the public, the legal community, and other stakeholders, as well as to scientists themselves. In this, as in the five chapters to follow, the Flora Danica project (1761–1883) and its history will illustrate and help us reflect on various pertinent aspects of the right to science. One of the largest botanical works ever published, *Flora Danica* covered the entire wild flora of the double monarchy of Denmark-Norway, Schleswig and Holstein, and the North Atlantic dependencies (the Faroe Islands, Iceland, and Greenland). Mostly forgotten today outside of Denmark—and, if remembered at all, then mostly for lending its name to a porcelain dinner set produced in the 1790s that was decorated with motifs from the *Flora Danica* herbarium and likely intended as a gift for Russian empress Catherine II—this botanical work offers insight into the advancement of human knowledge and creativity that is useful for the present project.

Flora Danica was an enlightenment-era project and an early example of a work that involved both citizen science and science diplomacy. The intention behind it was utilitarian in two ways. First, the idea was to find out what kinds of plants were available for practical use in agriculture and medicine to boost the Danish economy. Second, and perhaps more interesting in our context, the hope of its various editors was to make available to people in the Danish realm valuable knowledge about the uses of plants and herbs. Framing my discussions on the relationship between science and culture in Chapter 2, the dissemination of science in Chapter 3, scientific freedom in Chapter 4, the global transfer of knowledge in Chapter 5, and the usefulness of human rights discourse in Chapter 6, the *Flora Danica* is used as an early illustration of and introduction to various aspects of the right to science.

Chapters 2 through 5 each discuss one of the four parts of ICESCR Article 15. Chapter 2 explores the right to science as a cultural human right and the implications of its being included with the right to culture and authors' rights in both UDHR Article 27 and ICESCR Article 15. The rights outlined in these two articles are often viewed as the core of cultural rights.[15] As some scholars see it, the drafters of the UDHR and the ICRSCR made a mistake by categorizing the right to science as a cultural right—a mistake that may in part explain why this right is underexplored and less well known than other human rights.[16] I argue, however, that the presentation of the right to science as a cultural right may be valuable in that its proximity to the right to participate in cultural life as well as to authors' rights may allow for ethical and human-centered deliberations to become more integral parts of the scientific endeavor.

Chapter 3 concerns the dissemination of science, outlined in ICESCR Article 15(2). As the drafters of the ICESCR knew, without dissemination, translation, or curation the right to science cannot be activated. The public can benefit from scientific progress only when scientific knowledge, data, and expertise are made universally accessible and when the benefits of the practice of science are universally shared.[17] Lifelong learning about scientific topics often takes place outside the classroom in informal environments such as science museums and libraries. By making the latest scientific discoveries and discussions available to everyone and involving the public in the histories, processes, and impacts of science, museums and libraries play central roles in the scientific training of citizens. I argue that the work of museums, which emerged more or less simultaneously with the modern system of scientific disciplines, provides a long-established gateway through which citizens can gain access to and activate their right to science. As places of research, education, and interactive engagement, science museums and libraries have to come to grips with and balance the rights of the public with those of scientists and curators on a daily basis. They therefore provide a good illustration of issues of relevance to the dissemination of science.

The focus of Chapter 4 is scientific freedom [Article 15(3)]. For most scholars, scientific freedom is crucial, the starting point for everything they do. In the most relevant human rights instruments as well as in General Comment No. 25 on science, the 2017 UNESCO Recommendation on Science and Scientific Researchers, and the special rapporteur's report concerning science from 2012, necessary restrictions on scientific freedom nevertheless feature prominently. In order to safeguard basic human rights principles such as human dignity and nondiscrimination, prior informed consent, confidentiality of data, and other kinds of protection from dual-use research are needed. From a human rights perspective, scientific responsibility is the flip side of scientific freedom. When it comes to dual use, the interests of the public may clash with those of scientists. The latter cherish their scientific freedom whereas the former call for important restrictions on that very freedom. On this issue, as on the topics of strategic funding and citizen science, we must acknowledge the clashing human rights interests involved in order to find equitable and just solutions.

Chapter 5 explores the right to science from the perspective of international cooperation and solidarity. The global aspect of the right to science encompasses both the importance of sharing the tangible results or products of scientific research around the world and the free flow of ideas and creativity. Here, it is especially the latter aspect that is discussed. The chapter's focus is on the historical, legal, and conceptual meanings of Article 15(4). This last part of Article 15 relies on, and makes sense only when read together with, the article's first three parts. In the current context of pandemics and climate change, global in scope like the interdependence of the world's economies, it is by no means the least significant. The issue of intellectual property and its relationship to authors' rights, outlined in Article 15(1), plays a prominent role when it comes to the global transfer of knowledge, as it does in many other right-to-science contexts. Touched on in previous chapters, too, intellectual property is explored, in this chapter, from the perspective of international cooperation and assistance, voiced as an imperative in the ICESCR and other major human rights instruments. The issues of scientific (mis)conduct and the importance of ethical standards in the conduct of science are also relevant in this context.

The final chapter, Chapter 6, rounds off the book by offering a discussion of the usefulness of the human rights discourse. Over the past many years, points of criticism have been raised against this discourse from various quarters. These include attacks on human rights as a western construct that has been imposed on the world under the pretense of universalism, as well as critiques of the lack of obligations and enforcement in practice of these rights. In this chapter, I defend what Mary Ann Glendon has called in the American context "rights talk" and what I have described elsewhere as "human rights talk."[18] I do so against the background of the right to science as discussed in the previous five chapters. I also look at the new Cold War–like atmosphere currently engulfing science that calls out for science diplomacy. Little known as it is, the right to science, I argue, clearly shows the promise of human rights as empowering rights.

When he was honored by the AAAS Science and Human Rights Coalition in Washington, D.C., in 2009, Richard Pierre Claude, the first author to write a monograph about science in the service of human rights, compared the progress made within science and human rights, respectively.[19] "Advances in science and technology: certainly so. Advances in human rights: far too few," he said.[20] This "differential pace between achievements in implementing human rights and advancements in science and technology demands our attention," he continued. "Certainly, we still face human rights violations of every variety, including genocides, torture as public policy, and extra-judicial killings, but we now live in a technologically wired global village brought to us through innovative applications of science and technology."[21]

Claude was right, in my opinion, when he pointed to the potential of science serving human rights in his 2002 book. And he was right again when he drew our attention, seven years later, to the very prominent roles science and technology play in our lives—and to how important it therefore is to be alert to the human rights implications of the "technologically wired global village" in which we all live today.

In Chapter 6 as in my Conclusion, I attempt to connect all the threads from the previous chapters into one coherent narrative. I end by suggesting a couple of additional perspectives that may serve as recommendations for further action and research.

A Brief Note on Method, Sources, and Acknowledgments

It is not only with regard to the *Flora Danica* project that I draw on my Danish background. Throughout the book, I use other examples from Danish history and culture to illustrate, introduce, and open up relevant issues for discussion. My reason for doing so is twofold. First, as a small country whose language and culture have limited range, Denmark provides a case study of the need for international cooperation and multilingualism in science. Second, I offer a lens through which to view right-to-science issues that is different from the Anglo-American one most often used in today's global academia.

Danish is a language spoken by only about 5.5 million people and is understood by perhaps a few more million other Nordics in Norway, Sweden, Greenland, the Faroe Islands, and Iceland. Apart, perhaps, from the concept of *hygge*—a quality of cosiness and comfortable conviviality that engenders a feeling of contentment or well-being, pronounced "hoo-guh"[22]—Danish culture is not much known outside Denmark itself. For the Danes, as for the members of any other small culture or community, linguistic diversity is an important issue. When it comes to the dissemination of scientific research, their right to science is not activated without translation—both from (typically) English into their native tongue and from the professional language used in scientific journals into a more popular and accessible discourse that they can relate to.

The very fact that the dissemination of global scientific results takes place at the local level may well keep Danish alive as a language of popular dissemination. But as an academic language, chances are that it will not survive. This is a justified worry of scholars in my country, especially in the humanities and social sciences, whose research concentrates on Denmark and the Nordic countries. They have been used to publishing their scholarship in Danish or Nordic journals, but feel increasing pressure to publish in more prestigious British or American journals. This means that they cannot write in their native tongue and that they have to present their scholarship in a different way. With a more international audience, they can no longer take any knowledge of the Nordic context for granted.

As English has become the global academic language, it has become easier for native speakers of English to use local examples and to assume that the rest of the world will be able to follow. This is not the case for someone writing about, say, Danish history or politics. There are plenty of things that have to be explained before writers can get to the arguments they want to make. So location matters. By using the *Flora Danica* project as a kind of narrative backbone, I have chosen a perspective from which to engage with the right to science that is unlike the one a native speaker of English would use. It is my hope that this offers a different, but still useful, kind of reading experience.

Many of the texts I use are primary sources such as the General Comments issued by the monitoring body of the ICESCR, UNESCO reports, and reports written by the UN special rapporteur in the field of cultural rights. I also engage with secondary texts written by fellow scholars, legal as well as cultural—and also botanical. Not being a botanist myself, I have relied heavily on the work of three Danish botanists who are specialists on the *Flora Danica* project. Ib Friis, Henning Knudsen, and Peter Wagner have all given very generously of their time to explain to a humanities scholar with no knowledge of their discipline the ins and outs of *Flora Danica* and its plants. For this, I am very grateful. I would also like to thank the Carlsberg Foundation whose Semper Ardens Monograph grant allowed me to finish this book.

This is not the first time that I have attempted to present legal technicalities with the help of and through cultural texts.[23] I juxtapose these cultural texts—in this case mostly *Flora Danica*-related texts—with the legal texts not so much in order to have them enter into dialogue with one another, but to have the cultural texts introduce and highlight important themes and issues. My hope is that this will be less challenging to nonlawyers—but also to lawyers who have never worked with the right to science. The potential of this particular human right deserves to be widely known.

CHAPTER 1

Setting the Scene

1. Everyone has the right freely to participate in the
   cultural life of the community, to enjoy the arts and to
   share in scientific advancement and its benefits.
2. Everyone has the right to the protection of the moral
   and material interests resulting from any scientific,
   literary or artistic production of which he is the author.
   —Article 27, Universal Declaration of Human Rights

There were two *Flora Danicas*. The first was published in 1648, the second—for my purposes more interesting—between 1761 and 1883. The latter will help set the scene for my investigation into the right to science. Both this right and the second *Flora Danica* are all about (Western) science, and both have their intellectual roots in the Enlightenment and its struggle against dogma, superstition, and ignorance.

The ambition to collect compendia of knowledge that kept the editors of the *Flora Danica* going for 122 years dates even further back. In the prologue to his 2004 book on the making of the famous French *Encyclopédie*, historian Philipp Blom tells us that the first encyclopedic endeavors we know of are cuneiform tablets in the archives of the kings of Mesopotamia, containing lists of all kinds of objects, such as different kinds of trees.[1] Later, the Greeks and the Romans also favored encyclopedic works. None of the Greek works has survived, but Pliny the Elder's *Natural History* was consulted as an authoritative source of worldly knowledge until well into the sixteenth century.[2] The Islamic world also undertook collective encyclopedic efforts, and a number of Arabic encyclopedias were translated into Latin—most famously by the seventh-century bishop and scholar Isidore of Seville, later to be patron saint of the Internet.[3] In terms of comprehensiveness, it was the Chinese, though, who published the largest encyclopedic enterprise in 1726: the *Gujin tushu jicheng* in 745 volumes.[4]

There was thus a rich global tradition on which the contributors to one of the best known of all encyclopedias, the French *Encyclopédie*, could draw for inspiration. Including Denis Diderot, Jean d'Alembert, Voltaire, Montesquieu, and Rousseau, these contributors succeeded in publishing twenty-seven volumes, containing 72,000 articles, between 1751 and 1772.[5] One entry in the *Encyclopédie* concerns the word "encyclopédie" itself. The meaning of this word is the "linking of areas of knowledge." And the goal of an encyclopédie, this entry informs us, "is to assemble all the knowledge scattered on the surface of the earth, to demonstrate the general system to the people with whom we live, & to transmit to the people who will come after us ... that our descendants, by becoming more learned, may become more virtuous & happier, & that we do not die without having merited being part of the human race."[6]

In the editorial policy from one of the early volumes, the editors outline their aims: "One will find in this work ... neither the *life of the Saints* ... nor the *genealogy of noble houses*, but the genealogy of sciences, more precious for those who can think ... not the *conquerors* who laid waste the earth, but the immortal geniuses who have enlightened it ... for this *Encyclopédie* owes everything to talent, nothing to titles, everything to the history of the human mind, and nothing to the vanity of men."[7] Responding to several critics who wanted the *Encyclopédie* to carry more articles on the Church and Church doctrine, the editorial team made it quite clear that theirs was an attempt to go up against received truths, especially religious truths. They did so at great personal cost. The *Encyclopédie* was perceived to be dangerous, inciting political as well as religious rebellion. The Catholic Church tried to stop it, and its contributors were threatened with imprisonment, even execution, by the political authorities, who were trying to forestall events that, as we now know, would lead to the French Revolution.[8]

The French *Encyclopédie* is often considered one of the beacons of the Enlightenment. Blom calls it "a triumph of reason in an unreasonable time" and "a turning-point in history: the moment when new ideas carried the day over bigotry and orthodoxy."[9] Others have shown less sympathy toward the Enlightenment and its views on humans and nature. Anna Tsing opens her 2015 monograph, *The Mushroom at the End of the World*, in the following way: "Ever since the Enlightenment, Western philosophers have shown us a Nature that is grand and universal but also passive and mechanical. Nature was a backdrop and resource for the moral intentionality of Man, which could tame and master Nature. It was left to fabulists, including non-Western and non-civilizational storytellers, to remind us of the lively activities of all beings, human and not human."[10] Though in different and in some respects contradictory ways, both Blom and Tsing bring forward important insights that are relevant to my exploration of the right to science. I shall therefore come back to them at various points throughout this book.

I start this chapter with a description of *Flora Danica*, the largest such atlas in the world and a Danish Enlightenment attempt at widening the knowledge of ordinary Danes about their world—in the spirit of the French *Encyclopédie*. Next, I outline what we know and do not know about the right to science as a human right—and what kind of scholarly and other work we need to undertake to realize its great potential. In this as in subsequent chapters, aspects of *Flora Danica* bookend thematic explorations of the right to science. It thereby forms a backdrop to or narrative arc for my argument that this particular human right has special relevance today because of its potential, when used responsibly, to bolster the freedom of science and culture for the benefit of scientists themselves, the public, and society in general.

Flora Danica

The word "flora" can mean two different things: the plants growing in a particular region and a book that deals with the plants to be found in this region.[11] One of the first floras in the latter sense, and an inspiration for the *Flora Danica*, was a book published in 1588 on the plants growing in the Harz area of Germany. Another source of inspiration was the more modest books of herbs without illustrations or descriptions that were published in other parts of Europe. These were much cheaper and enabled the less well-off to find medicinal plants they could either develop themselves or buy in their local pharmacy.

The publication of these books coincided with the rise of a general interest in botany as well as with the development of botanical gardens. Inspired by these trends, Danish king Christian IV (1577–1648) thought it was time for people in his kingdom to learn about plants and their use as medicine against illnesses. In 1645, he approached the head of the University of Copenhagen with a plan for producing a book about herbs and plants for the general reader. He required names to be in both Danish and Latin, and he wanted descriptions of the habitat and use of the plants:

> Inasmuch as God Almighty this Land/ amongst others/ with divers Herbs hath blessed/ by which the lowly Country-dweller without Means to seek the Advice of a Physician against Ailments and Diseases/. . . even as he harried be by suchlike/ readily is cured and restored: then We graciously solicit you/ speedily to prepare and commit to the Printing Press/ an Herbarium in Danish/ for the Weal of our lowliest Subjects/ in which the Herbs that common grow and native are to this Country/ with Names be listed in Danish and in Latin/ and the places where they are wont to grow: and to signify and propagate withal/ their best Uses against Ailments and Diseases/ for the Benefit of the common Man.[12]

As there were not yet any independent chairs in botany at the University of Copenhagen, it fell to a professor of anatomy, surgery, and botany, Simon Paulli, to carry out this work. Paulli finished his *Flora Danica* in 1648, only a few months after King Christian IV died. It was Christian's son, Frederik III (1609–1670), who received all 886 pages of text plus 92 pages of index and 372 plates of woodcuts of the plants.

Paulli's *Flora* became very popular and was much used over the next many years. At some point, however, demand arose for a more accurate depiction of Danish plants. German-born physician, botanist, economist, and statistician Georg Christian Oeder (1728–1791), who was to become the first editor and most important drafter of the second *Flora Danica*, wrote about Paulli's work, for example, that it was "to put it mildly, very incomplete."[13] From the planning stages to the collection of plants and their publication in copper engravings and on paper, the second *Flora Danica* was much more ambitious than Paulli's *Flora*.[14] The first of several editors, Oeder was interested in what we would today call citizen science. He knew that it would be impossible for him and his assistants to gather enough plants for the *Flora Danica*, the Kingdom of Denmark at this time encompassing in addition to Denmark itself also Schleswig-Holstein and Oldenburg-Delmenhorst (both part of what is today Northern Germany) and Norway, Iceland, the Faroe Islands, and Greenland. His idea was therefore to engage voluntary collaborators, especially gentlemen farmers and clergymen, but also students and correspondents throughout Denmark. As noted by Jean Anker in his history of the *Flora Danica*, "The voluntary collaborators throughout the realms should be able to give information about the names of the plants, together with a number of things about their use, especially as home-made remedies. For, thought Oeder, farmers were not afraid of making experiments, especially on themselves, which physicians could not carry out. People on the spot could likewise undertake far more thorough investigations of the different tracts than could Oeder himself; thus they could make considerable contributions to his Flora."[15] Although Denmark acquired colonies, forts, and trading posts in West Africa, the Caribbean, and the Indian subcontinent during the seventeenth and eighteenth centuries, *Flora Danica* was never intended to cover plants from these tropical parts of the realm.[16]

Oeder did not have an easy start in Denmark. Having studied medicine at the University of Göttingen under the famous anatomist, physiologist, and naturalist Albrecht von Haller, he was called to Copenhagen in 1751 on the recommendation of von Haller and others. As I describe in more detail in Chapter 3, the Danish king created a special position for Oeder as royal professor of botany at the botanical institutions that were founded at around this time. This position was outside the jurisdiction of the University of Copenhagen, which opposed the hiring of Oeder and failed him when he defended his thesis in 1752. These defenses were public, and the successful defense of a doctorate thesis was a necessary requirement for a professorship. The official reason given was his deficient Latin skills: Oeder "was able to read, but he did not speak [Latin] well."[17]

The king paid the new royal professor of botany out of the privy purse. In addition to overseeing the building of a new botanical garden in Copenhagen, the king wanted Oeder to establish a library for "the public benefit and use" of all present and future "lovers" of botany.[18] This new library for ordinary lending became Denmark's first public library, to be extended in due course to encompass all the natural sciences.[19] It opened in 1761—in rooms adjacent to Oeder's official residence at the Frederik's Hospital in Copenhagen so that he could take good care of the books.[20] The idea was, as Oeder informed the public in a notice in a local newspaper in January 1761, that the new library would be open to all botany lovers on Wednesdays and Saturdays between 10 A.M. and noon and that Oeder himself would offer public lectures on botany in those same rooms on Tuesday and Friday mornings between 11 A.M. and noon.[21] In order to promote a wider knowledge of botany as a practical science, Oeder furthermore intended to carry out experiments involving Danish drugs at the Frederiks Hospital in collaboration with senior medical physicians. Botany cannot, he repeatedly stated, "be of general use as long as it is a science only for the few."[22]

That same year, 1761, Oeder sent out subscription invitations to people in as well as outside Denmark. Four-page folders in Danish and French, these invitations included the first plate of the *Flora Danica*, depicting the cloudberry, and explained that the *Flora Danica* would appear in fascicles including sixty plates or engravings each. There would be one installment a year, and it was possible for subscribers to choose either a cheaper copy in black and white or one in color. The subscription rates would be reasonable, as the king would support the project financially.

The approved original, drawn and hand-colored by an artist, would be used as a muster plate and engraved on the copper plate. Illuminists would then copy the muster plate in the required number, and if a whole set of the *Flora Danica* was later ordered, the muster plates would be used again.[23]

Oeder's plan for the *Flora Danica* included five subprojects or items:

1. An introduction to botany in Danish.
2. A methodical list in pocketbook format of all plants in the country for use on expeditions.
3. A description of all the plants.
4. An illustrated work with copper engravings of all the plants.
5. A practical section in which the properties and uses of the plants are described.[24]

This plan turned out to be overly ambitious. Oeder did manage to publish ten fascicles between 1761 and 1771, one a year as promised (item 4), while he was the editor of the *Flora Danica*—quite an achievement. He also wrote an introduction to botany, issued in two parts in 1764 and 1766, respectively, in three parallel editions with titles and texts in Latin, German, and Danish (item 1). The promised methodical list appeared, though in a somewhat different form from the list of all Danish and Norwegian plants with their names in Danish, German, English, and French that Oeder had originally envisaged (item 2). The material was too large for a pocket-sized expedition checklist, so Oeder published an alphabetically arranged dictionary in 1769 of plant names in Danish, Swedish, Latin, German, English, and French, with the title pages and introductions in Latin, German, and Danish.

As for item 3, a description of all the plants, the king instructed Oeder that "it is Our most earnest will, that the descriptions of the plants occurring in it shall follow every fascicle."[25] Oeder never fully complied with the king's will, but instead published the plates without accompanying description.

Later, he attempted to write up descriptions. He intended to publish these in two parts but succeeded in completing only the first part before he became embroiled in the events surrounding the death of King Frederik V at the end of the 1760s and was removed from his post.[26]

The lack of a defined description of the plants was disappointing to the king and to members of the Danish government. But the fact that Oeder never succeeded in working on the practical part concerning the usefulness of the plants caused more dissatisfaction. Leading politicians in Denmark at the time saw the study of nature as a means of utilizing plants for the support of agriculture and trade. For them, nature was indeed, to quote Tsing again, "a backdrop and resource for the moral intentionality of Man, which could tame and master Nature." In 1757, a commission had been set up to explore ways in which future agricultural needs could be met in the Danish realms. In a rescript to members of the commission, which concerned the plans for a Danish flora, the king stated that he wanted them "to investigate minutely whether the most important part, that is the practical part of the botany, should not be commenced so soon as possible, and not be put off for another 20 years into the future, in which case the utility to Our subjects and their encouragement to take up the study of botany, and especially Our high purpose, would be delayed altogether too long."[27]

Oeder simply had too much on his hands to fulfill this wish. The problem was not that he was not himself interested in practical botany and economics. "It is my most serious intention to spread the knowledge of herbs and plants in order to make them more useful," he once wrote about his work with *Flora Danica*.[28] During his travels to find plants in Norway and other parts of the realm, Oeder gained a thorough knowledge of agriculture and trade and sent home reports on economic issues concerning, for example, the modernization of agriculture. He held enlightened views on the status of farmers and developed an interest in sociopolitical issues such as widow's pension funds and the structure of and recruitment into the army. He also explored public health issues like inoculation against infectious diseases such as smallpox, as I describe in Chapter 3.

The problem was rather that "publication of the part concerning the usefulness of the plants has been entrusted to a society of which I am a member," as Oeder wrote in his 1761 subscription invitation.[29] This society never met, and the practical section of the *Flora Danica* (item 5) never appeared, even though Oeder does seem to have worked on it.[30] The only text to appear on the plates themselves was "Flora Danica Tab," followed by a serial Roman numeral.[31] *Flora Danica* therefore had no real system of classification. Later editors did attempt to make up for the missing textual parts, but it was only four years after the completion in 1883 of *Flora Danica* that Johan Lange, the last editor of the work, published a *Nomenclator Flora Danica,* which contained alphabetical and systematic lists of all the plants featured.

Oeder and his successors as editors and publishers may not have fulfilled all the objectives of the authorities for the production of the *Flora Danica*, but it earned everyone involved a lot of prestige. The fact that it was completed— that different editors over 122 years succeeded in producing seventeen volumes in folio, issued in fifty-four installments or fascicles each containing sixty plates, with a total of 3,240 engraved copper plates, and that the whole project was supported financially throughout by the Danish Crown—made the *Flora Danica* a success story at the time. As the Danish botanist and chronologist of the *Flora Danica* Peter Wagner describes it, "*Flora Danica* was not to be defeated. It survived the Danish State bankruptcy of 1813, the loss of Norway and the Duchies of Schleswig and Holstein in 1814 and 1864 respectively, and the fall of the absolute monarchy in 1849."[32]

Free copies were sent not only to clergymen, grammar schools, and educated people around Denmark, but also to famous botanists and members of royal families throughout Europe. Carl Linnaeus in Uppsala received a copy, just as the Royal Horticultural Society in London was presented with a set as a gift.[33] By the early nineteenth century, the reputation of the *Flora Danica* had become sufficiently established that the German botanist and physician Kurt Sprengel expressed his regard for it "as an incomparable work" and a "botanical treasure chamber" whose good-quality illustrations "have contributed to give the work a first place among botanical aids."[34]

Another sign of the high regard in which the *Flora Danica* was held in professional circles was the fact that it made illustrated floras fashionable. In Austria, between 1773 and 1778, Nikolaus Joseph von Jacquin published a *Flora Austriaca*. In his foreword, he described it as "Austria's contribution to Flora Danica." When the English botanist William Curtis published his *Flora Londinensis* during the years 1777 to 1798, he also referred to the *Flora Danica*, "a work which is now carrying on in Denmark under the auspices of the king." And when Peter Simon Pallas issued his *Flora Rossica* in 1784, he wrote that "a great many" of the plant species to be found in Russia, Tartary, and Siberia "are the same as European plants that have already been illustrated in *Flora Danica* and *Flora Austriaca*."[35]

Today, the *Flora Danica* has largely been forgotten. It no longer carries scientific weight, and, outside of Denmark, virtually no one has heard about this Danish enlightenment adventure. When Danes hear the words *Flora Danica,* they mostly relate them to the dinner set, first produced in the 1790s and still sold on demand by the Royal Danish Porcelain Factory (today Royal Copenhagen), which is generally believed to have been intended as a present from the Danish king to the Russian empress Catherine II. I come back to the *Flora Danica* dinner set and to the ways in which *Flora Danica* has become a cultural artifact in addition to a scientific object in later chapters. Suffice it to say, in the words of another chronologist, that the *Flora Danica* herbarium was considered a "regal work" in its day. Not only did it owe "its very existence to the goodwill and munificence of the monarch, since the king himself at first paid the expenses from his private exchequer," but in terms of "its form and Contents," the *Flora Danica* was also "worthy of a king, or was like a king among books."[36]

The *Flora Danica* may be described as an early instrument of science diplomacy, both foreign and domestic. In terms of achieving foreign policy objectives, scientific research outcomes, as well as science itself as a process and way of communicating, were used as soft power to show off and to further specific political (power) interests. The domestic policy objective was twofold: to use scientific knowledge to advance Danish agriculture, thereby strengthening the economy, and to enlighten the Danes. Much like the editors of the French *Encyclopédie,* Oeder was driven by a sense of public duty and a wish to let the public benefit, materially as well as intellectually, from the latest scientific progress. It is this sense of science diplomacy to which I now turn, the right to science being about access to knowledge and, in the wording of the Universal Declaration of Human Rights (UDHR), the right of everyone "to share in scientific advancement and its benefits."

What We Know About the Right
to Science as a Human Right

Historical Background

The drafters of both the 1948 UDHR and the 1966 International Covenant on Economic, Social and Cultural Rights (ICESCR) considered the right to science to be a universal human right. It was enshrined in the fourth section of the UDHR (Articles 22 through 27), which was considered, at the time of its passage, to be the most groundbreaking part of the new declaration.[37] UDHR

Article 27 reads, "Everyone has the right freely to participate in the cultural life of the community, to enjoy the arts and to share in scientific advancement and its benefits."[38] In 1966, the wording of Article 27 was repeated almost verbatim in ICESCR, whose Article 15(1)(a) and (b) talk about the right of everyone "to participate in cultural life" and "to enjoy the benefits of scientific progress and its applications," respectively.[39]

For the drafters of the UDHR, economic, social, and cultural rights were "indispensable for [a person's] dignity and the free development of [their] personality"[40]—and the rights listed in Articles 23 through 27 were groundbreaking because they aimed at the realization of the development of one's self.[41] But it was especially during the drafting of the ICESCR that the importance of science as a human right was emphasized. The delegate representing UNESCO praised the right to science as a "coping-stone of the edifice of human rights" and saw it as "to a great extent the determining factor for the exercise by mankind as a whole of many other rights."[42] And as far as the delegate from the Philippines was concerned, the right to science "dealt with the noblest rights that could be attributed to an individual."[43]

As a result, the right to science figures as a fundamental human right in two of the three human rights instruments that are together referred to as the International Bill of Human Rights. It also plays an indirect role in the International Covenant on Civil and Political Rights (ICCPR), which forms the third part of the International Bill of Human Rights. Given the large effects that science and technology have in our daily lives, the right to science is often considered together with freedom of expression, including the freedoms to seek, receive, and impart information and ideas of all kinds, regardless of frontiers. The right to science is also often mentioned together with the right everyone has to take part in the conduct of public affairs, directly or through freely chosen representatives. Both of these rights are protected in the ICCPR—in Articles 19 and 25, respectively. In addition, the right of all peoples to self-determination, mentioned in Article 1 of both international covenants, is important in this context.[44]

The UDHR is a declaration, an ideal standard of rights held in common by nations, but it bears no force of law.[45] Work on codifying these rights into a legally binding entity began almost immediately. Geopolitical developments such as the Cold War and decolonization processes as well as differences in economic and social philosophies made it very difficult to reach agreement. Between 1949 and 1951, the Commission on Human Rights worked on a single draft covenant, which would include both economic, social, and cultural rights and civil and political rights. But in 1951, geopolitical pressure had grown so much that the General Assembly agreed to draft two separate covenants: the ICESCR and the ICCPR.

After almost twenty years of drafting debates, both covenants were adopted and opened for signature, ratification, and accession in 1966, and they entered into force in 1976. For the drafters, the question of implementation was very important and was intimately connected with the issue of whether there should be one or two covenants. They thought that the covenants should not just establish rights—as had the UDHR—but also make clear the responsibilities of states parties in relation to the implementation of these rights:

> The question of drafting one or two covenants was intimately related to the question of implementation. If no measures of implementation were to be formulated, it would make little difference whether one or two covenants were to be drafted. Generally speaking, civil and political rights were thought to be "legal" rights and could best be implemented by the creation of a good offices committee, while economic, social and cultural rights were thought to be "programme" rights and could best be implemented by the establishment of a system of periodic reports. Since the rights could be divided into two broad categories, which should be subject to different procedures of implementation, it would be both logical and convenient to formulate two separate covenants.[46]

Human rights were now divided into economic, social, and cultural rights on the one hand and civil and political rights on the other hand, and this division would influence how cultural rights, including right to science, came to be perceived. Scholars and practitioners around the world have since been unable to agree on the proper status not only of cultural rights but also of economic and social rights. The official UN position is that all human rights are "universal, indivisible and interdependent and interrelated," but even within the human rights community, anyone arguing in favor of the usefulness of the right to science has typically faced an uphill battle.[47] Many human rights scholars and lawyers, as well as many of the Western states (e.g., the United States), consider civil and political rights to be the most important part of the human rights spectrum. They view economic, social, and cultural rights as mere policy goals and as nonjusticiable.[48] Those who do show an interest in these rights tend to focus on economic and social rights only.[49] And if the issue of cultural rights comes up, people will typically have the right to participate in cultural life [Article 15(1)(a)] in mind. The right to science is thus disadvantaged in three ways: (1) it is the least well-known cultural right, (2) it is ignored by scholars who are interested in economic and social rights, and (3) it does not constitute a proper human right for those who perceive civil and political rights to be the core human rights.

Nonetheless, in addition to the UDHR and the ICESCR, a few normative frameworks related to the right to science have been developed. In 2005, the United Nations Educational, Scientific and Cultural Organization (UNESCO) adopted the Universal Declaration on Bioethics and Human Rights, which includes a provision on benefits sharing, just as the Recommendation on Science and Scientific Researchers (2017) and the Declaration of Ethical Principles in Relation to Climate Change (2017) both mention the right to science. The EU Charter on Fundamental Rights also includes freedom of the arts and sciences in Article 13: "The arts and scientific research shall be free of constraint. Academic freedom shall be respected."[50] And most important of all, the Committee on Economic, Social and Cultural Rights, the independent body monitoring the ICESCR, published General Comment No. 25 on science and economic, social, and cultural rights, Articles 15.1.b, 15.2, 15.3, and 15.4, in April 2020.[51]

Scholarly Consensus—Such as It Is at This Time

Scholars working on the right to science tend to agree that the protection in ICESCR Article 15(1)(b) of every person's right to benefit from science has three major elements: freedom of science, enjoyment of the benefits of scientific progress and its applications, and protection from adverse effects of science.[52] The first element, freedom of science, concerns respect for the freedom scientists have to conduct their research. "The freedom indispensable for scientific research and creative activity" is also protected under ICESCR Article 15(3). This shows, Klaus D. Beiter argues, that, "the crucial rationale for the protection of freedom of science must be seen to lie in the fact that it makes possible the discovery of the truth. Scientific truth may, in turn, yield beneficial applications of science."[53]

Science historian Naomi Oreskes talks about the consensus that may form among scientists at any given time. Oreskes' answer to the question raised in the title of her 2019 book, *Why Trust Science?*, involves the social character of scientific knowledge production, the way in which consensus is arrived at. "The key point is this," she writes:

> We have an overall basis for trust in the processes of scientific investigation, based on the social character of scientific inquiry and the collective critical evaluation of knowledge claims. And this is why, ex ante, we are justified in accepting the results of scientific analysis by scientists as likely to be warranted. . . . The critical scrutiny of scientific claims is . . . done . . . in communities of highly trained, credentialed experts, and through dedicated institutions such as peer-reviewed professional journals, specialist workshops, the annual meetings of scientific societies, and scientific assessments for policy purposes.[54]

Beiter and Oreskes both make important points about respect for scientific freedom and the collective critical evaluation of scientific claims. Unless scientific freedom exists, and unless the general public has trust in the scientific process, no scientific results will be produced from which the public and society in general can benefit. Other scholars would point to an oath-based system which might persuade the public that scientists take self-governance just as seriously as do doctors and lawyers, for example.[55]

The main beneficiaries of the second element of the protection of Article 15(1)(b) are the general public and other stakeholders. Enjoyment of the benefits of scientific progress and its applications relates to access to benefits, as well as to opportunities to participate and to contribute. A human rights approach automatically adds a focus on disadvantaged groups in societies. Applied to the right to science, "this requires a form of affirmative action, that is, specific investments in science and technologies likely to benefit those at the bottom of the economical [*sic*] and social scale."[56]

In recent years, it is the active participation of citizens in—rather than their somewhat more passive enjoyment of the benefits of—scientific activities that has especially interested scholars and activists. "Citizen science," encompassing everything from crowdsourcing and collecting information and data to hands-on activity in laboratories, refers to scientific contributions by ordinary people without special qualifications. Among the most popular projects are those in which citizen scientists monitor global biodiversity and help gather millions of observations every year.[57]

General Comment No. 25 on science directly acknowledges that citizen scientists constitute an important knowledge resource when it comes to the genesis of science in terms not only of making discoveries but also of understanding values and norms, and of disseminating scientific knowledge. It asks that states parties actively encourage citizen participation in scientific activities rather than prevent it: "The right enshrined in article 15(1)(b) encompasses not only a right to receive the benefits of the applications of scientific progress, but also a right to participate in scientific progress."[58] It should therefore be known and referred to as the right to participate in and to enjoy the benefits of scientific progress and its applications.[59]

Access to and participation of citizens in science is furthermore an essential element of democratic citizenship. Shaheed suggests in her report on the right to enjoy the benefits of science and its applications, for example, that the right to science should be read "in conjunction with, in particular the right of all peoples to self-determination and the right of everyone to take part in the conduct of public affairs."[60] Public consultations on scientific advances and their implications are necessary, she argues, for at least two reasons. First, such consultations can help make sure that science policy is, at least in part, formulated around concrete wishes of the public. This is especially important when we are talking about publicly funded research from which the public has a right to benefit. Second, a public check may help protect vulnerable populations such as indigenous peoples from "the negative consequences of scientific testing or applications on, in particular, their food security, health or environment."[61]

In a democratic society, scientific research can never be entirely free. It must always be conducted in a socially and ethically responsible manner.[62] This takes us to the third element of protection offered by ICESCR Article 15(1)(b): the protection from adverse effects of science. There is an inherent tension between this third element and the first element, freedom of science. Protecting the public from adverse effects of science puts a limit on the scientific freedom enjoyed by scientists. One way of putting it would be to say that scientific responsibility is the flip side of the scientific freedom promised to scientists by Article 15(3).

In a statement published in *Science* in 2017, the American Association for the Advancement of Science (AAAS) makes the relationship between scientific freedom and responsibility—and the connection of both with human rights—quite clear: "Scientific freedom and scientific responsibility are essential to the advancement of human knowledge for the benefit of all. Scientific freedom is the freedom to engage in scientific inquiry, pursue and apply knowledge, and communicate openly. This freedom is inextricably linked to and must be exercised in accordance with scientific responsibility. Scientific responsibility is the duty to conduct and apply science with integrity, in the interest of humanity, in a spirit of stewardship for the environment, and with respect for human rights."[63]

At a time when many worry about the dual use of science and technology, this third element highlighting scientific responsibility gains new importance. "Dual use" refers to technology and items that can be used both for civilian or peaceful aims and for military aims. The right to science provides a tool for talking about this important issue—and for making sure that the effort if not to discover the truth, then to reach a scientific consensus on issues of vital importance to society such as climate change and sustainable development, takes places in a legally, ethically, and culturally sound way.

Human rights are not just a set of legal rights. They are also a set of ethical principles, providing "a value-added component to research ethics, as an internationally recognized framework of legal protection for the subjects of scientific research."[64] Working with Article 15 means working toward preventing the use or misuse of science and technology for purposes contrary to the enjoyment of human rights and contrary to human rights principles such as universality, dignity, nondiscrimination, equality of treatment, and inclusion. Once fully developed, Article 15 can be a gift to the scientific community as well as to the public and society in general. It offers scientists an opportunity to become engaged in establishing guidelines that can enhance the ethical practice of science, thereby promoting the common good—and to do so before such guidelines will be supplied by public authorities in the shape of administrative regulations.

What We Do Not Know About the Right to Science as a Human Right

Today, "the challenge that human rights scholars, practitioners, and intergovernmental organizations face is how to fulfil the promises of the UDHR and the ICESCR."[65] How do we activate and realize the right to science? One place to start is to draw attention to its existence. In general, the human rights community as well as the public, politicians, and other stakeholders are not aware of the promise inherent in this particular human right. Neither are scientists themselves.

"[Much] can be achieved," as the AAAS Science and Human Rights Coalition notes, "when the science, engineering and health communities embrace human rights as an area suitable for and deserving of robust inquiry, and become an influential voice in the defense of human rights."[66] In this regard, the coalition's own efforts over the past decade to elicit the perspectives of scientists and health professionals are instructive.[67]

The AAAS Science and Human Rights Coalition contributed two reports outlining the perspectives of scientists about the right to science. The report *Defining the Right to Enjoy the Benefits of Scientific Progress and Its Applications: American Scientists' Perspectives* was published in 2013.[68] As the title implies, this report focused on the attitudes of, and research done by, American scientists. In order to add a more global perspective, the report *Giving Meaning to the Right to Science: A Global and Multidisciplinary Approach* was published in 2017.[69] Like the 2009 *Venice Statement on the Right to Enjoy the Benefits of Scientific Progress and Its Applications*, one of the few other recent attempts at clarifying the right to science, the two AAAS reports try to mobilize science and scientists to advance human rights, especially the right to science.[70]

The study *The Right to Science—From Principle to Practice and the Role of National Science Academies*, represents the third stage of the efforts by the AAAS Science and Human Rights Coalition. This study suggests that national academies of science within states parties can play an important role in the implementation of the right to science. Serving as intermediaries to identify key priorities concerning this right and then framing these within their respective national contexts, academies of science could come up with locally relevant recommendations for how their governments might fulfill their obligations under Article 15.[71]

In 2017, inspired by the work of the AAAS Science and Human Rights Coalition, Lauren Segal asked young scientists to "describe how applications of knowledge in your field (information, methodologies, services, and/or products) could support civil, political, economic, social, or cultural rights." She received responses from scientists around the world representing a variety of fields. While none of these responses concerned the right to science directly, several of them mentioned rights overlapping with this human right. Two young scientists related their work on the fortification of foods to prevent malnutrition and on genetically modified crops capable of resisting harsh climates to "rights to food," while others listed their research on longer-lasting vaccines and wearable personal health trackers as of importance to people's "right to health." One young scientist associated their scholarship on forensic anthropology to identify victims of genocide with the "right to be remembered," while another chose the "right to culture" as relevant to a description of their work on neural underpinnings of variations in empathic response to members of different races or ethnicities.[72]

If mobilization of the scientific community is slowly happening, there is not much attention to the right to science by the general public or among policymakers yet—even though the stakes are pretty high. As the drafters of the UDHR knew, the right to science, along with the right to education and other cultural rights, is important for the realization of human dreams and aspirations on the personal, individual level. But society also has a stake in these rights being implemented. An educated, well-informed, and enlightened citizenry constitutes, they thought, the best defense against hatred and bigotry—and for democracy.[73] The drafters of the UDHR wrote against the background of the Second World War, but their sentiments would have been familiar to the editors of both the French *Encyclopédie* and the *Flora Danica*, who also perceived scientific enlightenment to be a bulwark against dangerous misconceptions.

The importance for society and for democratic citizenship of the right to science is the reason behind Shaheed's wish to dedicate one of her ten reports to this particular human right, as we saw. Thinking back on being the first mandate holder in the field of cultural rights, Shaheed mentioned "the paucity of previous work on cultural rights" as one of her main problems when attempting to explicate the scope and normative content of the right to science. This paucity was "reflected in the frequent comment prior to my mandate, that cultural rights were a generally underdeveloped area in comparison to other human rights."[74] At the time, there were not many civil society groups working on cultural rights, including the right to science, and human rights and other scholars did not seem to show much interest in this right either. When Shaheed's successor as special rapporteur, Karima Bennoune, took over in 2015, the situation had already changed somewhat. By then, Bennoune did have a fair amount of scholarship and a fair number of civil society groups to support her work.[75]

Much still needs to be done, though—in terms not only of making the right to science better known but also of developing its normative content in more detail and with greater clarity. It is only when such clarity is arrived at that this crucial human right can be activated and be used in practice. The publication of the General Comment on science marks a great step forward.

Of Alternative Facts, Post-Truth, and Fake News

What may get the scientific community and the general public interested in the right to science is the feeling many have of living in a so-called post-truth world, where scientific facts commonly agreed upon by scientists are routinely reduced to "fake news" or "alternative facts." Some argue that we should stop talking about posttruth so as not to turn it into a self-fulfilling prophecy.[76] Within the past couple of years, however, "the urgency of sorting truth from falsehood—information from disinformation—has exploded into public consciousness."[77] This merits our attention in the present context— especially since "sorting truth from falsehood" often involves not just scientific but also cultural controversy.[78]

People are generally not aware of how hard it is to reach scientific consensus.[79] Neither do they know much about the mechanisms through which this happens, if it happens: the gathering and then sharing of data through scientific societies, specialist conferences, and publication in peer-reviewed scientific journals, which is followed by critical scrutiny by fellow scientists and an adjustment of views if the criticism voiced is deemed to be correct and fair. This is "a social process that rigorously vets claims," science being "simply the consensus of relevant experts on a matter after due consideration," and scientific facts being "claims about which scientists have come to agreement."[80]

As Oreskes reminds us, people tend to link science to its implications and utility. History shows us that utility has always played a part in political and economic support for science.[81] The *Flora Danica* provides a good example of how government (and the king's) support was linked to the value of plants in medicine, public health, and the economy in general. When people today question scientific consensus on issues such as vaccine safety or climate change, it is often because the cultural/religious/existential implications of pertinent scientific findings do not agree with their own values and philosophy of life. The conflict concerns not so much the scientific as the cultural (or existential) side of things: "On most of the scientific issues that are highly contested . . . there is a scientific consensus. What is lacking is cultural acceptance by parties who have found a way to challenge the science. This is the source of the contestation, not conflicting positions within the scientific community."[82] There may, at one and the same time, be both consensus *within* the scientific community and plenty of conflict *outside* this community—or rather, between certain groups of people and the scientific community. I agree with Oreskes that the best response on the part of the scientific community is not to be defensive and refuse to talk about norms and values. Scientists are taught that science is supposed to be objective ("let the data speak for themselves") and that what they believe is not relevant.[83] Epistemologically speaking, such neutrality may well be incontestable, but it does "not work to permit communication and build bonds of trust."[84] In addition to emphasizing how difficult scientific consensus is to come by and explaining how the scientific process works, scientists would do well to admit that they, too, have values and that such values matter. "Would you trust a person who has no values?" Oreskes asks.[85] Striving for value neutrality "is more or less the equivalent of saying that [science] has no values—at least none other than knowledge production—and this can elide into the implication that *scientists* have no values. Clearly, this is not the case, but scientists' reluctance to discuss their values can give the impression that their values are problematic—and need to be hidden—or perhaps that they have no values at all."[86] This is good neither for scientists nor for the public and other stakeholders who increasingly rely in their daily lives on science and its products. Why not let people know that scientists are deeply dedicated to making a contribution to their particular field of science in order to cure diseases, create a better world, or make the environment ecologically safer? After all, these are values that can immediately be understood as they are shared by many nonscientists too.

Using a human rights discourse automatically adds an ethical dimension, as we saw above. The responses Lauren Segal received from young scientists when she asked them how they see their scientific work supporting human rights is a case in point. A human rights discourse, especially a right to science discourse, also makes the relationship between scientific freedom and responsibility quite clear—a relationship that, in light of public fears of dual use, scientists need to acknowledge more openly than they currently do. After all, "if the history of science teaches anything, it is humility."[87] Scientists should be prepared to explain the basis of their claims and be open about the possibility that they are wrong and may fail—or that (their) science has not yet come as far as they would like it to have come.

The definition or explanation of what is or is not science—and what characterizes a good scientist—is one that is heavily discussed. One recent attempt at clarification is the 2017 UNESCO Recommendation on Science and Scientific Researchers in which the term "scientific researchers" "signifies those persons responsible for and engaged in research and development," and the term "the sciences" "signifies a complex of knowledge, fact and hypothesis, in which the theoretical element is capable of being validated in the short or long term, and to that extent includes the sciences concerned with social facts and phenomena."[88] The word "science" itself

> the enterprise whereby humankind, acting individually or in small or large groups, makes an organized attempt, by means of the objective study of observed phenomena and its validation through sharing of findings and data and through peer review, to discover and master the chain of causalities, relations or interactions; brings together in a coordinated form subsystems of knowledge by means of systematic reflection and conceptualization; and thereby furnishes itself with the opportunity of using, to its own advantage, understanding of the processes and phenomena occurring in nature and society.[89]

These definitions are repeated verbatim in the General Comment on science.[90] Along with Oreskes' work, the General Comment and the UNESCO recommendation give us some useful working definitions. Over the next chapters, as I approach discussions concerning the relationship between science and culture, scientific dissemination, scientific freedom, and the global transfer of knowledge, I will come back to these definitions and their implications for the right to science.

Conclusion

In 1771, when the *Flora Danica* was in its tenth year of publication, Voltaire sent a letter, or rather an epistle, to Danish king Christian VII (1749–1808), praising him for abolishing censorship in September 1770:

> O King, virtue of honor, though despotic born! . . .
> *Thinking you allowed, you gave back its right;*
> *Physics, Song, Sermon, Novel, History, Singing Games,*
> *Anyone can write anything, and pipe whatever they want.*[91]

Voltaire could equally have paid tribute to Johann Friedrich Struensee. Christian VII had become king of Denmark–Norway in 1766 at the tender age of sixteen. Danish historians speculate that he probably suffered from schizophrenia.[92] Between 1770 and 1772 it was his German-born court physician, Struensee (1737–1772), who served as the kingdom's de facto ruler. Struensee was inspired by the French Enlightenment and philosophers such as Voltaire, and he tried to introduce reforms intended to relieve poverty and oppression. One of these concerned censorship, another the modernization of the legal system, including the abolishment of torture. These reforms were considered by some to be too progressive, and when Struensee also openly conducted an affair with Queen Caroline Mathilde (1751–1775), he was arrested under the false accusation of having plotted to kill the king and was executed.

The brief Struensee era is one of the most remarkable chapters in Danish and Scandinavian history—and the inspiration for a number of creative works such as the 1999 novel by Swedish writer Per Olov Enquist, *The Royal Physician's Visit*, and the 2012 Danish movie *A Royal Affair*.[93] For our purposes, Struensee is interesting because Georg Christian Oeder was associated with him and carried out his work on the *Flora Danica* in the spirit of the Struensee reforms. When Struensee was executed, people who had supported or just been close to him lost influence. Oeder was one of these and he was reassigned, first to a post as chief administrative officer in Norway, and then eventually as governor of Oldenburg, in what is today Northern Germany.[94]

Struensee's execution in 1772 led to a two-year hiatus in *Flora Danica*-related activities. Publication resumed with the appointment of Oeder's successor, Otto Friedrich Müller (1730–1784), as editor and publisher in 1774. What ultimately saved the *Flora Danica* was its fame and its beauty, especially the high quality of its copper plates and illustrations.[95] Oeder knew that the financial success of an undertaking such as the *Flora Danica* hinged on its aesthetic appeal and that visual accuracy was of the essence with respect to the scientific and practical merits of the project. Unless plant drawings were precise, academic and amateur botanists could not benefit from the results of Oeder's work, and the general public would have a limited chance of recognizing these plants. On both counts, the usefulness of the project would suffer. Oeder therefore invested a lot of energy in finding the very best craftsmen. When it became difficult to keep up with the demand for the colored version of the *Flora Danica*, he even went so far as to open the School of Illumination for Women in Copenhagen, as we shall see in Chapter 2.

From the beginning, scientific excellence thus went hand in hand with visual and artistic excellence, just as plans for utilizing knowledge about nature for economic and agricultural gain were part and parcel of enlightened wishes for giving larger parts of the population access to the latest scientific knowledge. In this latter regard, both Tsing's criticism of the Enlightenment view of nature as but something to be tamed and mastered and Blom's emphasis on

Enlightenment thinking as "a triumph of reason in an unreasonable time" are relevant. The *Flora Danica* was fueled by economic necessity—but this should not make us forget that Oeder and the men who supported him were inspired by the kind of thinking that informed the French *Encyclopédie*. Theirs was an honest wish to produce knowledge for the benefit of the public.

Science and/as culture, and science in the service of both the material and the intellectual betterment of humankind: in the history of the *Flora Danica* we find a number of the topics reflected or illustrated that are today discussed with relation to the human right to science. On a day of general discussion, held in Geneva in October 2018 by the UN Committee on Economic, Social and Cultural Rights on a draft General Comment on Article 15 of the ICESCR, remarks made by both invited speakers and members of the committee made it clear how much opinions differ on the purpose and usefulness of the right to science. For some, the word "applications" is the most important, the value of "the right to enjoy the benefits of scientific progress and its applications" lying in the practical transfer of technology from the global north (as chief producer, at least for now, of the knowledge behind such technology) to other parts of the world. In this rather narrow reading of ICESCR Article 15, the right to science is all about access to technology and the material products of science.[96]

But broader interpretations of the right to science were also voiced that day in Geneva. In my own contribution, I argued that Article 15 is more than just material benefits; it is about reaffirming science as a part of democracy, and about highlighting the value of evidence-based inquiry and policymaking.[97] When the General Comment was finally published in April 2020, the Committee on Economic, Social and Cultural Rights made it clear that the right to science is about both the material benefits and the development of knowledge: "The term, "benefits" refers first to the material results of the application of scientific research, such as vaccination, fertilizers, technological instruments and the like. Secondly, benefits refer to the scientific knowledge and information directly deriving from scientific activity, as science provides benefits through the development and dissemination of the knowledge itself. Finally, benefits refer also to science's role in forming critical and responsible citizens who are able to participate fully in democratic society."[98] Oeder would have understood both the narrow and the more sweeping arguments for the importance of the right to science. So would the drafters of the International Bill of Human Rights, who discussed this very issue. Then, as today, there was a certain tension between those who saw the right to science as a means to combat prejudice of various kinds and those for whom the importance of this particular human right primarily concerns science and technology transfer.[99]

The name *Flora Danica* has been employed in many different contexts over the years, both commercial and noncommercial. What started out as an Enlightenment idea about collecting scientific knowledge about plants in the Kingdom of Denmark, including Norway, Iceland, the Faroe Islands, Greenland, and parts of what is today Northern Germany, has undergone quite a cultural journey. It is to this cultural journey and to science as a cultural human right that we shall now turn.

CHAPTER 2

The Right to Science as
a Cultural Human Right

The States Parties to the present Covenant recognize the
right of everyone:
   (a) To take part in cultural life;
   (b) To enjoy the benefits of scientific progress and its
applications;
   (c) To benefit from the protection of the moral and
material interests resulting from any scientific, literary or
artistic production of which he is the author.
          —Article 15(1), International Covenant on
          Economic, Social and Cultural Rights

This chapter explores the right to science as a cultural human right. During
the drafting process of both the UDHR and the ICESCR, the nature of the
right to science was discussed, but the reasons for classifying it as a cultural
human right—rather than as, say, an economic or a social human right—
were never made clear. Some scholars see the categorization of the right to
science as a cultural human right as a mistake on the part of the drafters of the
UDHR and the ICRSCR—one that may help explain why this right has lain
dormant for so long.[1] I see it in a more positive light: A cultural human rights
perspective enhances the use of science as an instrument for human benefit
by encouraging ethical deliberations and by prioritizing human rights and
human-centered considerations above commercial interests.

This chapter is divided into three parts. In the first, I analyze *Flora Danica*
as a cultural icon—a fine example not only of the interaction between science
and culture but also of the importance of culture and art for the growing

significance of science. Scientifically speaking, as we saw in Chapter 1, *Flora Danica* has lost much of its influence. Scholarly articles and books are still published, but these typically concern the historical significance of a particular aspect of the work.[2] The name *Flora Danica* also from time to time crops up in a scientific context,[3] and it could well be that biologists and others interested in issues related to, for example, climate change or environmental changes may increasingly want to use *Flora Danica* as a signpost to understand today's flora.

But it is primarily its use as a *cultural* artifact that has kept *Flora Danica* alive. This should come as no surprise. It was the high quality and beauty of its copper plates and drawings that originally helped make the work famous—and that ensured its continuation when Georg Christian Oeder, its first and most famous editor, became a pawn in the political infighting in Denmark after the execution of Johann Friedrich Struensee, court physician to King Christian VII.[4] Most famous of all is the Flora Danica dinner set, of which some 1,530 original pieces are preserved. But this set is only one part of the remarkable cultural afterlife of *Flora Danica*. A number of artists, in and beyond Denmark, have been inspired by what originated as a serious scientific (and economic) attempt at enlightenment, just as designers have been motivated by the work to create Flora Danica scarves, bags, and jewelry.[5]

The next two parts of this chapter investigate the implications of the right to science being included with the right to culture and authors' rights in both UDHR Article 27 and ICESCR Article 15. Along with the right to education, the rights outlined in these two articles are normally seen as the core of cultural rights.[6] During the drafting and adoption of the American Declaration of the Rights and Duties of Man (1948), which preceded the UDHR and inspired its drafters, the right to science was mentioned as a stand-alone right, separate from the right to culture. It was only later that it became part of the broader right to "take part in the cultural life of the community, to enjoy the arts, and to participate in the benefits that result from intellectual progress."[7]

I discuss whether the connections between the three parts of ICESCR Article 15(1) can pave the way for ethical and human-centered deliberations becoming a more integral part of the scientific endeavor. A part of this discussion concerns the question of authors' rights, as outlined in Article 15(1) (c), and their relation to the rights to culture and science. Under investigation is whether, and if so how, viewing the three parts of Article 15(1) together allows for a (better) discussion of the integration of academic research into the market. This integration sometimes demands a price on the credibility of scientists and potentially stifles a much-needed open and honest dialogue between scientists and the public concerning access to knowledge, intellectual property (IP), financial interests, and corporate affiliations.

In the third and final part of the chapter, I look at science as a practice that is socially and culturally situated in time and space. A cultural rights approach invites us to place science in its societal context and to open up scientific issues to different, but no less relevant, lines of inquiry. Every scientist is raised in a particular cultural context that will inevitably have an effect on how they see the world and their research, no matter how much they try to exclude normative issues of any kind. "Because [science] is a discipline given to objective fact-finding addressed by hypotheses and using inductive methodologies, it appears by its reliance on empiricism to exclude all consideration of value-laden issues. Whether this view stands up to scrutiny or not, it remains obvious that science is a discipline pursued by human beings," as Richard Pierre Claude, the first scholar to write a monograph on the right to science, once put it.[8]

Like everyone else, moreover, scientists may make mistakes and take their work in a direction that they should not. People must be protected against the negative consequences of such research and its applications to, for example, their food, health, security, and environment. Along with the other rights outlined in Article 15 ICESCR, the right to science embodies principles that are intended to inform the conduct of science.[9] As a *cultural* human right, I argue, it links scientific freedom to scientific responsibility and "adds a legal and moral dimension to a range of fundamental issues, including scientific freedom, funding, and policy, as well as access to data, materials, and knowledge."[10]

Providing a tool for defending responsible science and the freedom to think, question, and share scientific ideas, the right to science is important for researchers but just as much for society.

Flora Danica as a Cultural Artifact and Icon

From the very beginning, drawings of specimens as true to life as possible were crucial for understanding and recording the potential uses of plants, whether economic, medicinal, or scientific. "Where accurate figures are given," stated British naturalist George Edwards in 1758, "much pains may be spared in verbal descriptions."[11] Scientists and naturalists participated in a number of the big voyages of discovery and exploration around the world that were undertaken by European colonial powers from the seventeenth century onward. Some, like Alexander von Humboldt, were skilled artists themselves, but most of them were unable to make accurate drawings of the plant specimens they found and therefore chose to travel in the company of artists.[12]

Flora Danica also involved a good deal of traveling. Plants were collected from around the entire Danish realm—with the exception of what was then known as Tranquebar (today Tharangambadi, on India's south coast), which became Denmark's first colony in 1620 and was sold to Britain in 1845, and the Danish West Indies, which were sold to the United States in 1917.[13] Plants from the tropical colonies were never intended to be included in *Flora Danica*.[14] None of the available sources mention why plants from the colonies outside Europe were not going to be a part of the *Flora Danica* project. Sending out expeditions to the tropical realms may have been too expensive, we may speculate. Besides, the idea was to describe plants that could be of immediate use to Danes (see Chapter 5).

Oeder and his successors as editors of *Flora Danica* instead spent a fair amount of time traveling to Norway and other parts of the kingdom. On these journeys, artists came along whose drawings served a taxonomic function—identifying, classifying, and naming plants—while at the same time providing useful information for policymakers back in Denmark. Botanical expertise was a highly valued and valuable form of practical knowledge.[15] Natural history art not only served as an aid for scientists when they were traveling abroad; recognizing and selecting plants could be "the difference between healing or poisoning oneself and others."[16]

In botanical drawing, a plant is often shown at various stages of development, in bud, in blossom, and in fruit, just as the internal structure is shown at greater magnification. This method owes a lot to the introduction in the mid-eighteenth century of Carl Linnaeus's classification system based on the sexual characteristics of plants.[17] First described in *Systema Naturae* in 1735, Linnaeus's taxonomic system "offered the artist a method of revealing the required facts about a species for classification purposes."[18] As discussed in Chapter 4, Oeder did not see eye to eye with Linnaeus on the issue of taxonomy. Even so, the explosion in botanical art in the late eighteenth and early nineteenth centuries, of which *Flora Danica* was a part, did show the direct impact Carl Linnaeus had on the discipline of botany.[19]

It was Germany, Oeder's birth country, which fostered some of the first scientific artists of the eighteenth century. One of the most famous was Georg

Dionysius Ehret (1708–1770), who, after leaving Germany for good, traveled to Holland only eventually to end up in England, where Joseph Banks
employed him to illustrate the plants that he had collected on his various
travels.[20] Ehret became so famous that he could publish scientific works in
his own name; Linnaeus even named a new plant after him, *Ehretia tinifolia.*[21]
When attempting to find good and competent illustrators for *Flora Danica*,
Ehret was one of the artists Oeder turned to.

In France, too, a number of flower painters and botanical illustrators were
employed at the Jardin du Roi. Created in 1626, the Jardin des Plantes, as it
came to be called after the French Revolution, remains the main botanical
garden in Paris to this day. Before the Revolution, a number of flower painters
at the garden became famous in their own right. One of them was Madeleine
Françoise Basseporte (1701–1787), who seems to have been a personal favorite of King Louis XV (1710–1774).[22] Serving for nearly fifty years as an official
botanical illustrator at the Jardin du Roi, Basseporte was one of the female
"enthusiastic naturalists and artists from all layers of society" for whom "the
Enlightenment opened the way."[23] Female illustrators were to play an important role in the early years of *Flora Danica*. Due to its immediate success,
Oeder had trouble finding enough illustrators. More people wanted the colored than the black-and-white version, and this surprised Oeder, who had to
ask the Danish king for extra funding to keep up with the large demand for
"illuminated" copies.[24] Part of the solution turned out to be the opening of the
School of Illumination for Women.

Oeder was very well aware of the importance of botanical artists and
thought that a good illustration was much to be preferred to incorrect and
inadequate descriptions.[25] "If, since the invention of the art of printing text
and engravings jointly," as he wrote in his invitation letter to subscribe to
Flora Danica in 1761, "one had provided more good drawings instead of
incomplete descriptions, then botanists would have been saved much of the
sour work necessary to identify and apply synonyms, and there would have
been less reason to complain about the amount of redundant names; then
the essential knowledge of plants would have become more widespread than
now, and one would have progressed more with knowledge about the uses of
plants than is presently the case."[26] Sadly, Oeder's views on the significance
of drawings were not shared by the Danish policymakers who hired him to
develop a botanical garden and library in Copenhagen and to work on publishing a Danish flora. A few years before the first volume of *Flora Danica*
was published, Oeder suggested three possible candidates for him to work

with. All three were German artists who knew their own worth and asked for a salary almost as high as the one Oeder himself received. A panel of professors at the newly opened Royal Danish Academy of Painting, Sculpture and Architecture (today the Royal Danish Academy of Fine Arts) was asked to assess the three candidates and to recommend one of their own students if they thought one was good enough.[27]

The answers given by the professors revealed that they had no idea what to expect of a botanical artist. Seriously underestimating the artistic skills involved, they thought that the academy could easily produce a candidate since "this kind of drawing is one of the poorest, and demands more industry than intelligence."[28] They also pointed to a particular student at the academy who claimed to need only a couple of weeks of intensive training to be able to deliver what Oeder wanted—and who was willing to undertake this job at a considerably lower salary than the one asked for by the German candidates. But at this point, it was obvious even to Oeder's employers that, time being of the essence as work on *Flora Danica* could not proceed without a skilled botanical artist, Oeder would be ill served with someone uneducated in and with no knowledge of botanical drawing. They therefore encouraged Oeder to travel to England, Holland, France, and Germany and to search for a skilled botanical artist. Aiming high, Oeder asked Ehret, who politely declined, so Oeder had to keep searching. He finally succeeded in contracting with copper plate engraver Michael Roessler from Nürnberg, Germany, and his son Martin.[29]

The cooperation with the Roesslers was very productive, but father and son could not keep up with the demand. Something had to be done and in 1762, the School of Illumination for Women opened. Two "teacher mothers," who were themselves competent colorists, were hired to take care of the education of the young women employed, many of whom came from poor families. The idea of educating young women of limited means may have come to Oeder during a visit to Paris. There, he saw how Madeleine Basseporte educated poor young women free of charge so that they could secure a living as flower painters.[30] The school seems to have worked well—but even so, like the Roesslers, the young women had difficulty keeping up with the rising demand for colored sets of *Flora Danica*. This left Oeder no choice but to outsource part of the work to competent colorists in Germany. Danish policymakers were none too keen on *Flora Danica* being produced outside of Denmark, though, so they allowed Oeder to bring a few more competent foreign artists to Copenhagen. One of these was Johann Christoph Bayer

(1738–1812), who came to Denmark from Germany in 1768 to draw originals for the *Flora Danica* copper plates. Eight years later, Bayer transferred to the Royal Copenhagen Porcelain Manufactory (now Royal Copenhagen), where he was put in charge of the coloring of the *Flora Danica* dinner set.[31]

The Dinner Set and the Continued Use
of the Flora Danica Name

Of the original 1802 Flora Danica dinner set pieces, about 1,530 have survived and are used for joyful events in the royal household, such as birthdays, weddings, and banquets. Some of these pieces are on permanent display at Christiansborg Palace and Rosenborg Castle in Copenhagen.[32]

The idea for a dinner set that would use and copy the *Flora Danica* drawings of plants may have come from Johan Theodor Holmskjold (1731–1793), director of the Royal Danish Porcelain Manufactory from its foundation in 1775 until his death.[33] Production took place over twelve years, but not much is known about it.[34] A set was ordered in 1790, and it is generally believed, but has never been proved, that this set was intended as a present from the Danish king to the Russian empress Catherine II. The empress was a passionate collector of porcelain, and flower paintings on porcelain were common at the time. Catherine died in 1796, however, and the set stayed in Denmark.[35]

In 1797, a new order was placed by the royal household, but production was stopped by the king himself in 1802 before all the ordered pieces were finished. The factory offered to make up the deficiencies, but this happened only much later.[36] When Princess Alexandra, the daughter of Danish king Christian IX, was married to the heir apparent to the British throne (later King Edward VII) in 1863, the couple was given a Flora Danica dinner set as a wedding present. Unlike the pieces produced before 1802 that were intended to be as scientifically correct—and as close to the original drawings—as possible, the set given to Alexandra and Edward consisted of new plates and items that were simply pleasing to the eye. Aesthetic considerations had at this point become more important than scientific accuracy.[37]

The Flora Danica dinner set has remained in production and is today still made on demand according to the same guidelines as the original dinner set: "The tradition and craft of painting Flora Danica is developed, refined and passed down through generations. The art of painting Flora Danica is a rare talent; there are less than 20 Flora Danica artists in the entire world. The craft requires years of training, an in-depth understanding of botany as well as patience, as one must slowly cultivate and let the flowers 'bloom' on the fine porcelain. In addition, a deep understanding of colour combination is needed, as the colour palette of Flora Danica is highly complex."[38] When the current Danish crown prince, Frederik, was married to Mary Donaldson in 2004, they too were presented with a complete Flora Danica dinner set. Theirs was a gift from the Danish people, who had collected money for it throughout the country.

Flora Danica is clearly alive and well—and has become a constant source of inspiration for artists, designers, and other creative people. This has happened mostly in the Danish context, but there are examples of foreign artists also being stimulated by this "science in bloom."[39] One of these is Belgian artist Pierre Alechinsky, who, in a series of drawings from the 1980s and 1990s called Flora Danica, used discarded prints from the work as background. The original plants are visible in all the drawings. Mushrooms have been turned into eyes or function as the decoration of a woman's dress, for example; grass is growing as hair on top of a human head, while a detail of the plant makes up the tongue—or, cabbage forms the hair of a person.[40]

More recently, in 2013, the Natural History Museum of Denmark ran the exhibition Flora Danica: Science and Art. The idea was to (re)tell the story of Flora Danica by presenting a number of original, hand-colored prints side by side with interpretations of the old work by contemporary Danish artists.[41] A special exhibition, Flora Danica Zoom, which focused on craft and textiles, was added to the main exhibition.[42] One of the botanists involved in the exhibition, Henning Knudsen, published a magnificent work on Flora Danica the following year, in which he not only told the story of the original work and its plants, but also reprinted some of the most beautiful drawings and prints.[43]

In 2012, Danish designer Anja Vang Kragh designed Flora, a new dinner set for Royal Copenhagen "beautifully honouring the legacy of Flora Danica," and carrying "the historic floral motifs into the future."[44] Likewise, Danish accessory design company Flora D designs scarves and bags with Flora Danica motives.[45] There is also the Flora Danica jewelry line (today called Flora Danica Jewelry of Copenhagen). Founded in Copenhagen in 1953, the company originally used real botanical materials as a basis for their jewelry, coating them with either silver or gold. In 1992, the company was asked by the royal household to produce a special memorial brooch, designed by Queen Margrethe II personally.[46] Both Flora Danica Jewelry of Copenhagen and Georg Jensen carry a jewelry line focused on the daisy, Queen Margrethe's favorite flower.[47] For this, as well as for other reasons, the *Flora Danica* daisy is particularly interesting, and I shall come back to it later.

Finally, the renewed interest in wild plants in response to climate change and environmental concerns in general has given *Flora Danica* new significance. In 2004, a group of Nordic chefs, including a couple of now-famous Danish chefs, drew up their New Nordic Kitchen Manifesto. Aiming to promote Nordic cuisine, this manifesto emphasizes the importance of plants and produce that are unique to Nordic climates, landscapes, and waters. It calls for the combination of good taste with modern knowledge of health and well-being, the development of new ways of utilizing traditional Nordic plants, as well as the blending of the best Nordic recipes and culinary traditions with foreign food influences.[48]

A number of the wild, edible plants used in New Nordic Kitchen recipes are listed in *Flora Danica*. One prominent example is elder (*Sambucus nigra*), whose flowers and berries have been utilized for centuries, as we shall see in later chapters. The veneration by these Danish artists and cultural entrepreneurs for the original *Flora Danica* and the wish to be true to and carry into the future the classic *Flora Danica* prints are remarkable.

The Right to Science: Paving the Way for Ethical and Human-Centered Deliberations

If botany meant, for the young illuminators of *Flora Danica*, a chance to earn a living, then for their better-off sisters it opened the door somewhat to a scientific education. In the eighteenth century, the language and arguments of botany became "inextricably implicated in arguments about women's intellectual and moral faculties and their general social status."[49] Jean-Jacques Rosseau promoted botany as a feminine pursuit and offered guidance and instruction to women, for example. Closer to nature than men are, he thought, women have a special affinity for botanical exploration.[50]

That Rousseau and others saw botany as one of the very few sciences appropriate for women to study is not surprising, "given that botany grew out of an alliance of herbals, healing and gardening, areas in which women had long been active."[51] For the famous French philosopher as well as for many others at the time, exciting women's interest in botany did, however, present one major problem: Carl Linnaeus's taxonomy detailing the sexuality of plants might stir in botanizing women an unhealthy curiosity about their own sexuality. This not only could destroy the nostalgic view of flowers as symbols of innocence, but it might also lead to all kinds of sexual anxieties that could jeopardize the relationship between men and women. Botany could therefore be both sexy and dangerous.[52]

These paternalistic fears notwithstanding, it was Enlightenment botany, "moving away from the particularized knowledge of the old herbals and embracing the universal systematizing of Linnaeus," that gave women access to botanical knowledge for the first time.[53] At my own university, the University of Copenhagen, women currently make up by far the majority of biology students. Yet, in every region of the world, including my own, women researchers remain the minority in the STEM fields as a whole.[54] UNESCO data from 2014 to 2016 show that female enrollment is especially low in information and communications technology; globally, women represent only 3 percent. In the natural sciences, mathematics, and statistics, that figure is 5 percent.[55] Gender equality is accordingly one of UNESCO's global priorities.[56] Goal 5 of the seventeen sustainable development goals of the 2030 UN Agenda for Sustainable Development also concerns gender equality and the empowerment of all women and girls.[57]

UNESCO and the UN have furthermore increased their focus on gender equality by requesting that the UN special rapporteur in the field of cultural rights integrate a gender perspective into their work.[58] In her report on gender, first mandate holder Farida Shaheed makes her most forceful statement for the importance of cultural rights. Emphasizing the universality of cultural rights and the principles of nondiscrimination and equality that underlie all human rights, Shaheed argues that constant cultural negotiation is good for democracy. We should keep questioning relativist discourses and cultural practices that are defended in the name of culture yet have a negative impact on women. By doing so, we can help stimulate discussion that may lead to the destruction of harmful hierarchies and a reorientation of culture toward embracing human rights.[59]

Naomi Oreskes has made a similar point. It is sometimes argued that the work of feminist scientists and historians of science may obstruct the advance of science. Yet, feminist criticism of science has never been intended to obstruct or deconstruct science and scientific methods, she maintains, but rather to widen the discussion by taking into account all the various perspectives that women bring to the table. By voicing concerns that may be different from those of men, these feminists have done us all a favor. The world of science, as indeed the democracies of which science forms an important part, is better off when confronted with and having to relate to diverse experiences and worldviews. This is how science works, by first testing and then either incorporating or refuting such theories and views. The consensus reached in the end is much the stronger for the different lines of inquiry being raised and debated along the way by women, but also by minority and indigenous groups: "A community with diverse values is more likely to identify and challenge prejudicial beliefs embedded in, or masquerading as, scientific theory."[60]

The UN special rapporteur's mandate also calls for the integration of a disabilities perspective into the mandate holder's work.[61] Both Shaheed and her successor as mandate holder, Karima Bennoune, have worked a great deal on gender issues, and by their own admission, the disabilities perspective has not received nearly as much attention as that on gender.[62] In our context, what matters is not whether Shaheed and Bennoune have devoted enough attention to persons with disabilities. Instead, what matters is that a cultural human rights perspective—and here, more specifically, a right to science perspective—paves the way for ethical and human-centered deliberations. Special attention is called for to make sure that the needs of underrepresented groups are heard—and that hurtful beliefs are exposed and do not enter science as perceived truths. We see this reflected in CESCR General Comment No. 25 on Science, in which both women and people with disabilities are mentioned as groups that need special protection.[63]

This seems to have implications for the ways in which scientists should be educated and trained: to basic education in science should be added perspectives on the possible implications of scientific research for human beings and the environment in which we all live. According to the 2017 UNESCO Recommendation on Science and Scientific Researchers, for example, the initial education of scientific researchers ought to involve, among other things, "the ability to review a problem or situation in perspective and in proportion, *with all its human implications*" (emphasis added).[64]

Connecting Authors' Rights to the Rights to Culture and Science

The 2017 UNESCO Recommendation on Science and Scientific Researchers is an important standard-setting instrument in that it both lists the goals and value systems by which science operates and underlines the need for protection of these goals if science and scientists are to prosper.[65]

One recommendation is "to fully respect the intellectual property rights of individual researchers."[66] IP rights, including patents, copyright, industrial design rights, and trademarks, are mentioned in other human rights instruments. For example, in the European Charter of Fundamental Rights, which became legally binding on the EU with the entry into force of the Treaty of Lisbon in December 2009, the second part of Article 17, concerning the right to property, says that "intellectual property shall be protected."[67]

The modern justification for IP rights is economic and utilitarian. Sometimes referred to in terms of "incentives-for-creation," the main idea is that there would be insufficient motivation to invent, create, and build commercial goodwill without IP law. IP protection is usually given for a finite period of time, typically twenty years in the case of patents. For copyright, in EU countries, the exclusive rights to a work lapse seventy years after the death of the author or originator.[68] Anyone who wants to make use of a patented or copyrighted work has to get a license to do so. This typically costs money and is subject to the discretion of the patent or copyright holder.

In the context of the right to science, ensuring the rights of scientists and other producers of scientific knowledge could come into conflict with the need to make this knowledge accessible. A systematic review of the extant literature shows that an overwhelming majority of articles dealing with the right to science touches on the topic of access—access on the part of the general public, but also of researchers, to all parts of science, from the necessary education all the way to the data, knowledge, and applications that arise from scientific inquiry.[69] Today, public access to the latest scientific discoveries and discussions primarily involves digital media. Significant divides exist; current trends toward mandatory open access journals and repositories notwithstanding, not everyone around the world can afford access to the Internet.[70]

The tension between IP protection and access to technology and information is mentioned both in the 2017 UNESCO recommendation and in General Comment No. 25 on Science. Acknowledging that scientists benefit from IP as producers of scientific knowledge on the one hand, General Comment No. 25 notes how IP can have a negative effect on the advancement of science and people's access to its benefits, on the other.[71] Access to knowledge, including open access, is just as important for scientists themselves, as they depend completely on access to the knowledge produced by their colleagues around the world.[72] General Comment No. 25 also notes how a large part of scientific research is today carried out within companies and by nonstate actors.[73] As a consequence, levels of IP protection have been rising in recent decades, and commercial interests now play a much larger role in science funding and priority setting than when the ICESCR was drafted and discussed.[74]

The World Trade Organization (WTO) Agreement on Trade-Related Aspects of Intellectual Property Rights, which came into force in 1995, imposed on all contracting members of the WTO minimum standards of protection for several areas of IP and so became "the turning point for international IP law."[75] It changed existing national patent laws at a time when many countries had no IP laws or, as in the case of developing countries, excluded patents on, for example, pharmaceuticals.[76] One result has been that trade rights and commercial interests have been given priority over human interests as outlined in the international human rights system. This has affected the world of science. As funding increasingly comes from private or commercial sources, commercialization has threatened to exclude the needs and interests of the less wealthy when research priorities are set.

What does all of this mean with respect to ICESCR Article 15—and, more specifically, what is the relationship between IP rights and authors' rights? Does the right outlined in Article 15(1)(c), "to benefit from the protection of the moral and material interests resulting from any scientific, literary or artistic production of which he is the author," mean that IP rights are human rights? Scholars with an interest in the right to science have debated this question for a long time. Authors' rights concern the reputation and integrity of creators—the rights of attribution and of objecting to any mutilation or other distortion of one's work.[77] The gist of relevant UN documents and reports such as those written by the special rapporteur in the field of cultural rights is that IP rights are not human rights.[78] "There is no human right to patent protection," writes Farida Shaheed. "Where patent rights and human rights are in conflict, human rights must prevail." Patents (and copyright) must never "extend so far as to interfere with individuals' dignity and well-being."[79]

Comparing IP regimes with the wording of Article 15(1)(c) is deceptive, Shaheed argues, because cultural rights, like all other human rights, are inalienable (that is, not transferable to others), whereas IP is a monopoly someone has for a limited period—typically twenty years in the case of patents—that mostly protects business and corporate interests. After this time, the IP-protected item or knowledge falls into the public domain and can be used free of charge by others.[80] Following the reasoning of CESCR General Comment No. 17, Shaheed ties the material interests of creators (and authors) to the enjoyment of an adequate standard of living.[81] Like any other type of work, scientific (and artistic) work can be protected by measures beyond IP such as minimum wage protections, collective bargaining power, social security guarantees, and budgetary support for science (and the arts).[82] General Comment No. 25 on science likewise emphasizes the importance of state-provided financial support to counter the most negative effects of IP. Such support can be given either at the national level or, if the state does not have the necessary means, through international cooperation.[83]

As Shaheed's reports, the 2017 UNESCO recommendation, and the 2020 general comment on science show, a cultural rights perspective changes the focus of analysis. It reframes legal discourses in such a way that they no longer automatically protect IP over individual rights and social values. As Aurora Plomer, one of the few IP experts working on the relationship between IP and human rights, nicely sums it up, "IP rights are time and space bound." By contrast, Plomer says, "human rights enunciate moral ideals based on the primacy and dignity of each human being. They are universal and hold irrespective of place and time. They articulate the spheres of civil, political, social, economic and cultural protection which are required for the full realization and development of each human being's personality. It is this ethos of human self-realization which animates the right to science in Article 27 UDHR and Article 15 ICESCR."[84] In this way, cultural rights provide a tool for prioritizing human rights and human-centered considerations above commercial interests. This is especially the case when all three parts of ICESCR Article 15(1) play together.

Science in Its Societal and Cultural Contexts

A cultural rights approach allows us to place science in its societal context and to open up scientific issues to different, but no less relevant, lines of inquiry. As some of the most influential UDHR drafters saw it, cultural life includes science—and science, much like the arts, is a practice: Tara Smith notes that "by the time the Drafting Committee of the Commission on Human Rights considered the right to science, the French delegate, Rene Cassin, affirmed the point of view that John Humphrey had conveyed in the UN Secretariat's draft, that 'cultural life included science.' This was a sentiment echoed by the Chinese delegate, Mr. Chang who observed that 'the word "cultural" could have two meanings; it could refer either to the practice of science and the arts, or the ethnical origin of a community.'"[85] Following in the footsteps of John Humphrey, René Cassin, and Peng Chun Chang, Farida Shaheed has maintained in all her reports that cultural rights relate to the pursuit of knowledge and understanding as well as to human creativity in both science and culture. Learning to appreciate the connection between science and culture was one of the most useful lessons she learned as a special rapporteur.[86]

As the cultural rights mandate was developed, a number of states voiced the fear that cultural rights would lead to or even encourage cultural relativism. This fear may well have been caused by the wording of the mandate, which requests that the mandate holder "study the relationship between cultural rights and cultural diversity."[87] Some scholars from the natural and technological sciences are also concerned.[88] Parts of the humanities being, as they see it, prone to cultural relativism, these scholars fear that such considerations may have a negative influence on the more empirically oriented, evidence-based sciences.

Working with the right to science as a cultural right, however, is not an invitation to inflict cultural relativism on science. Like all other human rights, the right to science is a universal right. This has been emphasized in both the 2017 UNESCO Recommendation on Science and the 2020 General Comment on Science, as well as in almost every special rapporteur report since 2009. Karima Bennoune even used the occasion of the seventieth anniversary of the UDHR to reexamine the cultural rights approach to the universality of human rights. In her 2018 report, *Universality, Cultural Diversity and Cultural Rights*, she sees universality and cultural diversity as mutually reinforcing principles, cultural rights providing a supportive environment for cultural diversity.[89]

Grounded in the struggle against all forms of discrimination, "universality is the cornerstone of human rights law."[90] It is today threatened by fundamentalist actors who have a background in and focus on religion, Bennoune argues, and by extremist forces with other bases from all parts of the political spectrum. Both fundamentalist and extremist forces routinely make use of religious, cultural, or ethnic arguments, and in the process they restrict scientific and artistic freedom: "We face a worldwide struggle," says Bennoune, "to defend intellectual freedom and the rationality on which it is based. Moreover, at the heart of the fundamentalist and extremist paradigms are reflections of equality and universality of human rights, making unwavering defence of those principles the touchstone of the human rights response. . . . Arts, education, science and culture are among the best ways to fight fundamentalism and extremism. They are not luxuries, but critical to creating alternatives"[91] Once again, women's rights are directly involved. Targeting women and claiming to defend religion or cultural traditions, both paradigms reject any substantive equality between men and women and aim to suppress women's cultural rights and women's voices. Often, the right to education and to gain knowledge, especially scientific knowledge, is directly involved. The attempt to change the content of education by removing sex education or scientific topics that question fundamentalist religious views concerning gender is a good example.[92]

In defense of universality, Bennoune maintains, echoing Shaheed, a human rights-based response to fundamentalist and extremist paradigms must be "fully gender sensitive. . . . Women's human rights, including cultural rights, are an essential part of the fight against fundamentalism and extremism, without which it cannot succeed."[93] Such a gender-based response to fundamentalist and extremist attacks on universality is a good example of how using a right to science approach allows us to tackle fundamentally important questions by combining objective factfinding and inductive scientific methodologies with qualitative research sensitive to cultural issues.

Though not voiced in right to science terms, Paul Anastas's call for a "green chemistry" that will incorporate sustainability and the protection of the planet and its inhabitants into chemical design and engineering provides a second example.[94] In a 2020 interview, Anastas explains how he sees an expansion of the definition of performance as well as a reliance on a wider set of disciplinary skills as necessary for today's chemists: "In traditional approaches to design—whether it be cars, electronics, plastics, or paints—we focus on fairly narrow definitions of performance. And historically, we have designed products that have often worked well for the narrow function, e.g., lead paint, but caused unintended consequences. In order to avoid this in the future we need to think more broadly about not just narrow, specific performance but also the effect that a product or process may have on the environment, the worker or the consumer. This will require a more diverse disciplinary skill set than only [that of] mechanical engineers for example."[95]

The realization on the part of scientists of the need for inclusion of sustainability considerations in their research is one we will see much more of in years to come, I think. As is the case with Anastas and his co-workers, such a realization may come from personal conviction—or it may be forced on the world of science by popular political demand, voiced, for example, in the form of an right to science argument.

Science and Ethics

The issue of dual-use scientific research, and the role of ethics in science in general, offers a third example of the advantage of a cultural rights approach to science. As we saw in Chapter 1, the protection in ICESCR Article 15(1) (b) of every person's right to benefit from science has three major elements: freedom of science, enjoyment of the benefits of scientific progress and its applications, and protection from adverse effects of science.[96] When it comes to the third element, protecting society and individuals against negative effects of science, the issue of dual use, in both a narrow and a broad sense, is very relevant. The impacts of some scientific advances depend crucially on how they are implemented and used. Nuclear physics, for example, may be used to build both power plants and warheads. This phenomenon, in which one application of a scientific concept or technology leads to beneficial and another to deleterious consequences, is known as the dual-use problem. Tellingly, the phenomenon was first used to describe rocket science during the cold war: the technology that could launch an astronaut into space is the same technology that launches intercontinental ballistic missiles. Dual use since then has referred broadly to any technology that can be used in the pursuit of two separate aims, but usually it is used in a narrow sense to describe technology with a civilian use and a military use.

Even before the cold war had really started, the dropping of atomic bombs on Hiroshima and Nagasaki in August 1945 had made the need for ethics in scientific research obvious. At the conference in London in November 1945 organized to establish the United Nations, British minister of education Ellen Wilkinson used the opportunity to explain that as a consequence of what had happened at Hiroshima and Nagasaki, the British delegation would like to propose that science be included in what was originally meant to be the United Nations Educational and Cultural Organization—or UNECO. What helped put the S into UNESCO was the realization of how important it is to make scientists ponder and take responsibility for the possible consequences of their research.[97] One way to do so was to have them be more closely "linked with the humanities," Wilkinson argued: "In these days, when we are all wondering, perhaps apprehensively, what the scientists will do to us next, it is important that they should be linked closely with the humanities and should feel that they have a responsibility to mankind for the result of their labours. I do not believe any scientists will have survived the world catastrophe, who will still say that they are utterly uninterested in the social implications of their discoveries."[98]

The Second World War and the way in which Nazi doctors had abused Jewish and other patients in various concentration camps—ostensibly to further scientific progress—also played a role in a new ethical awareness. Since then, the use of technology has become omnipresent in our lives, and the concern about dual use has widened accordingly. A 2017 AAAS statement, linking scientific freedom to scientific responsibility, reflects this concern, as we saw in Chapter 1. So do the numerous ethics councils that have been created in the past few decades around the world, as well as the popularity of the codes of ethics for business, designed to help establish values-based rules for employees to follow. In the world of scientific research, the completion of ethical issues tables now being demanded in research funding applications is especially noteworthy. For all activities funded by the European Union, for example, "ethics is an integral part of research from beginning to end, and ethical compliance is seen as pivotal to achieve real research excellence."[99]

One of the communities that has been directly affected by both the beneficial and the deleterious consequences of science is people with disabilities.[100] Science and the products of science are critical to their inclusion in society and their ability to function in society. Yet, as modern technology has allowed members of this community to take charge of their own lives and to live independently, people with disabilities have also been unwittingly subjected to scientific experimentation with potentially serious health consequences. For example, writes Valerie J. Bradley, "in 1949, at Fernald State School (previously the Massachusetts School for the Feeble Minded), 74 boys aged 10 to 17 were recruited to join the 'Science Club.' They were given special privileges but were also given oatmeal for breakfast with milk laced with radiative tracers. In another experiment, some of the boys were injected with radioactive calcium. These experiments were approved by the Atomic Energy Commission. Ironically, some of the boys enthusiastically joined the club thinking that the scientists would expose the abuse that went on at the facility."[101] It is not difficult to see that the conduct of the scientists involved in these and other studies, who took advantage of vulnerable and cognitively compromised individuals, is appalling and should be condemned on ethical grounds. There are other cases, though, that are less straightforward. To come up with solutions to such problems requires the ability to work across disciplines and practices—from the natural and life sciences to the humanities. I shall come back to the issue of dual use in later chapters.

Conclusion

In a (now famous) lecture at the University of Cambridge in 1959, C. P. Snow debated the gap—and the proper balance—between the technological and natural sciences on the one hand and the humanities on the other. He called this gap "the two cultures" divide and argued that the British educational system had overrewarded the humanities at the expense of scientific and engineering education.[102] The two-cultures gap is still with us, but today it is hardly the natural and technological sciences that are at a disadvantage. The power relationship between this part of the academy and the humanities has changed completely. Many humanities scholars are concerned as they see their disciplines losing status and funding—even as core parts of the humanities such as philosophy and ethics are increasingly in demand. Taking the right to science seriously as a cultural human right may not bridge the two-cultures gap completely, but it could help make it smaller and increase the willingness to ask questions and to see pathways from a variety of disciplines and angles.

Like many other academies around the world, the Royal Danish Academy of Sciences and Letters, which was founded in 1742 and of which several *Flora Danica* editors were members, endorses and supports the understanding of interdisciplinary research. This is reflected in the publications of the academy. The 2010 volume of essays, *The Foreign as a Powerful Historical Force in Denmark After 1742* is one fine example.[103] In chapters written by scholars of the humanities and of the social, natural, and technical sciences, the volume tells of the many ways in which foreign scientists have had an impact on all parts of Danish science and, indeed, on Danish life in general.

The 2010 publication was designed as a *Festschrift* honoring the seventieth birthday of Queen Margrethe II, who is the protector of the academy. The queen's nickname is Daisy, after a flower that has been linked to various figures in history. *Marguerite* being the French word for daisy, it was the emblem of Queen Margaret of Anjou (1430–1482), for example.[104] Members of the Danish Royal Academy therefore chose to display the *Flora Danica* daisy on the front cover of the *Festschrift*. In her introduction, editor Marita

Akhøj Nielsen explains, with clear reference to Oeder and his *Flora Danica* project, that "our point of departure is the marguerite on the cover, a beautiful representative of the Danish flora. . . . The systematic exploration of all plants in the country during the age of enlightenment was largely conducted by foreigners."[105]

The marguerite or oxeye daisy is portrayed in plate 994 of *Flora Danica*, volume 6, published in 1790. Scientifically known as *Leucanthemum vulgare*, it is a globetrotting wildflower that spreads easily and today grows in many parts of the world. It has meant many different things to many people, has played a role in herbal medicine, and has been a versatile ingredient in the kitchen. In Nordic mythology, the daisy was dedicated to the goddess of love and fertility, Freya. Its Latin name, *Bellis perennis*, has been connected to the Roman myth of the nymph Belides, who, in order to escape the advances of Vertumnus, god of seasons and gardens, turned herself into a daisy.[106]

Like other natural symbols, the daisy is common currency in tradition and folklore. The flower is supposed to have clairvoyant powers, and people have associated it with innocence, youth, and vitality. Children (used to) wear daisy chains, garlands created from daisy flowers. The common daisy crops up in several of William Shakespeare's plays and poems. In *Hamlet*, it is one of the flowers of the drowning Ophelia's "fantastic garland," and in the narrative poem *The Rape of Lucrece*, the daisy is used as a metaphor for virginal whiteness and the arrival of spring.[107] Shakespeare was not the only writer to deploy daisy metaphors. In his fairy tales, Hans Christian Andersen also put traditional knowledge about plants, including the daisy, to creative use. In "The Butterfly" (1861), we hear, for example, how "the Butterfly flew down to the Daisy, whom the French call 'Marguerite.' They know she can tell fortunes. This is the way it's done: the lovers pluck off the little petals one by one, asking questions about each other, 'Does he love me from the heart? A little? A lot? Or loves me not at all?'—or something like that; everyone asks in his own language."[108]

Beyond paying tribute to the queen, the choice made by the Royal Academy to take the both ordinary and extraordinary daisy as a starting point for an interdisciplinary discussion of the influence of the foreign on the local was an excellent one. "The essence of Flora Danica is the interplay of science and art," announced Hanne Strager, curator and head of exhibitions, as the 2013 exhibition opened in Copenhagen.[109] It is this very interplay that is reflected in ICESCR Article 15, which aims to further human creativity and learning, in terms of not only science and technology, but also artistic expression.[110] And it is this interplay of science, art, and culture that helps open up science to lines of inquiry emphasizing ethical, human, and non-commercial interests.

What culture, the arts, and the humanities can do is to help popularize and make people engage with science in all sorts of creative ways. It is to the dissemination of science that I will now turn.

CHAPTER 3

The Dissemination of Science

The steps to be taken by the States Parties to the present
Covenant to achieve the full realization of this right
shall include those necessary for the conservation, the
development and the diffusion of science and culture.
—Article 15(2), International Covenant on Economic,
Social and Cultural Rights

Without dissemination, translation, or curation there will be no right to science. The public can truly benefit from scientific progress only when scientific knowledge, data, and expertise are made universally accessible and when the benefits of the practice of science are universally shared.[1] And, unless scientists have venues beyond the scientific world itself in which their scientific breakthroughs can be popularized and harnessed, the chance of showing the usefulness and importance of their basic research to the public is limited. The drafters of the UDHR in 1948 and the ICESCR in 1966 understood this. ICESCR Article 15(2) made it obligatory on states parties to the ICESCR to include, in their steps toward achieving the full realization of the right to enjoy the benefits of scientific progress and its applications, "those necessary for the conservation, the development and the diffusion of science and culture."[2]

In this chapter, science museums and libraries provide the point of departure for an exploration of ICESCR Article 15(2). These institutions play a crucial role in preserving and disseminating science. My emphasis is on the latter—on the way in which by making the latest scientific discoveries and discussions available to everyone, science libraries and museums are go-to sources of information, both print and digital, on today's critical global issues. Moving away from didactic instruction toward immersion, interactive collaboration, and the thoughtful articulation of scientific ideas and values, they have undergone great changes in the past few years. Today, science libraries and museums offer places for conversations concerning how science may contribute more effectively to major issues in the public domain and promote human flourishing, and how the freedom to think and share, but also question, scientific ideas may be defended.

By making use of what we would today call citizen science and by officiating the creation of the first public library in Copenhagen, Georg Christian Oeder was also aware, as we saw in Chapter 1, of the potential for scientific institutions to be enablers of science. The first step toward the publication of *Flora Danica* was a resolution, signed by King Frederik V (1723–1766) on October 3, 1752, which concerned the construction of a library for "the public benefit and use" of all current and future "lovers" of botany.[3] The first part of the chapter takes a closer look at this library, which opened in 1761, the year Oeder sent out the first subscription invitations to *Flora Danica*. Denmark's first public library, it would come to incorporate all the natural sciences, just as it would come to play a role in botany becoming a discipline in its own right rather than remaining a part of the medical sciences. Oeder's library was an enlightenment successor to the cabinets of curiosities of the sixteenth and seventeenth centuries collected by members of the merchant class and early practitioners of science—and a precursor to the modern science museum and library.

The second part of the chapter investigates the meaning of Article 15(2), legally and conceptually. For the drafters of the ICESCR it was important to move beyond the nice words and political proclamations of the UDHR to the more concrete obligations that these proclamations impose on states parties, as we saw in Chapter 1. But the ICESCR does not specify which steps states might or ought to take for "the conservation, the development and the diffusion of science and culture." CESCR General Comment No. 25 on science offers a few suggestions as to what Article 15(2) may require and encompass. I examine these and their implications for the discussion of, among other things, public funding for research, as well as for public participation in decision making. I will also look at historical sources such as the *Travaux Préparatoires* of the UDHR and the ICESCR to see whether they offer any guidance and clarification.

In the third and last part of the chapter, returning to the role of science libraries and museums, I explore science communication and citizen science and their implications vis-à-vis Article 15(2). As resources for public education and engagement in science research, these institutions wrestle with the question of how actively they should invite public discourse as well as policy on controversial or sensitive topics relating to, for example, climate change or human gene editing. Major challenges are international in scope, and science and its applications are part of the cause as well as the cure. How should this be reflected in exhibitions and other museum-related work? To what lengths should scientists go in disseminating their work? Should they increase their participation in traditional educational outreach programs, or should they interact more directly with the public in museum citizen science projects—perhaps even support the direct impact of citizen science on their research questions?

Drawing the public into scientific research, I argue in this chapter, is a positive development—for members of the public as well as for politicians and scientists themselves. From a human rights perspective, participation should include decision making. This means that the public has a right to participate in the formulation of science policy. But we should be attentive to the possibility of certain negative consequences. The majority of the population in any given country might want to prevent scientists from conducting research into politically and culturally sensitive topics such as gender or evolution, for example. Current tendencies toward allocating more money to strategic research initiatives, at the cost of free basic research, could also be furthered, which would be problematic for the freedom of inquiry and scientific freedom outlined in ICESCR Article 15(3).

Largely, the advantages outweigh the possible disadvantages. Citizen science empowers both the citizens and the scientists involved. It enables scientists to harness the power of people's curiosity, talent, and observations. It also offers possibilities for scientists to be heard and to impress on people, from the individual voter to their political representatives, the relevance of scientific research. This will be especially important as public funding decreases in many countries, and scientists must compete with a number of other actors for available funding.

As places of research, education, and interactive engagement, science museums and libraries have to come to grips with and balance the rights of the public with those of scientists and curators on a daily basis. They therefore provide a good illustration of the issues at stake.

A Copenhagen Library for the Public Benefit and Use of All Current and Future "Lovers" of Botany

Along with collecting, preserving, researching, and exhibiting artifacts, education has always been among the main tasks of science libraries and museums. How these tasks have been carried out and for what purposes have

varied greatly over time. The encyclopedic collections of the Renaissance cabinets of curiosities, emerging in the sixteenth century, were meant to astonish
and dazzle, for example, whereas the opulence of the Baroque museums of
the seventeenth and eighteenth centuries was intended as a demonstration
of status and wealth, colonial as well as national.[4] Combining in their exhibitions elements from both the Renaissance and Baroque museums, Enlightenment museums added an emphasis on motivating people to continue their
education.

The British Museum, established in 1753 as a public institution and free
to all to this day, was where visitors could see all kinds of natural and artificial
curiosities as well as read books and is often viewed as the Enlightenment
museum *par excellence*. Officially, its aim was very close to that of the French
Encyclopédie—namely, "to reexamine and question received ideas and values and explore new ideas in new ways. Through an empirical methodology,
guided by the light of reason, one could arrive at knowledge and universal
truths, providing liberation from ignorance and superstition that in turn
would lead to the progress, freedom and happiness of mankind."[5] A number
of scholars have since questioned this Enlightenment legacy of science and
reason. To Anna Tsing, as we saw in Chapter 1, Enlightenment thinkers had
a utilitarian view of nature and failed to see the interdependence of human
beings and nature. Historian Eric Hobsbawm criticized the Enlightenment
in a slightly different, though related, way when he wrote in 1997 that, "these
days the Enlightenment can be dismissed as anything from superficial and
intellectually naïve to a conspiracy of dead white men in periwigs to provide
the intellectual foundation for Western imperialism."[6]

The British Museum may well have been an inspiration for Georg Christian Oeder as he embarked on building a new botanical garden and establishing a public library in Copenhagen. Another was the growth of the natural
sciences at universities in other countries. In 1741, professor of history Hans
Gram (1685–1748), who served as rector of the University of Copenhagen for
a while and helped establish the Royal Danish Academy of Sciences and Letters in 1742, had suggested that it might be a good idea to start teaching modern topics such as economics and the practical use of plants. This, he argued,
was done to great effect and success in neighboring Sweden by Linnaeus and
others.[7] The attempt to have Oeder hired as a professor of botany at the University of Copenhagen was intended as a step in the same direction. Trained
by one of Europe's most famous medical scholars and botanists, Albrecht von
Haller, in Göttingen, Germany, Oeder was thought to be exactly the right man

for the job of instituting botany as an independent area of study. Not only were his scientific credentials immaculate, but Oeder had also displayed a strong interest in the economic and social aspects of applied botany and science.

Most of the powerful men working with and for the absolutist king at the time were large landholders so they had not just a professional but also a personal investment in the latest knowledge on agriculture, gardening, and forestry.[8] They had successfully persuaded the king to create for Oeder a royal professorship of botany, paid for directly by the royal household, having refused to give up when the University of Copenhagen failed Oeder's doctoral dissertation, as we saw in Chapter 1. The oral defense of the dissertation took place on February 16, 1752. Oeder had come to Denmark only toward the end of the previous year, so he did not have much time to get ready. The event attracted quite a crowd. "I had a large auditorium, and there were famous people present, their Excellencies von Holstein, Count Raben, Baron Korf, Russion envoyé and Baron Gyldenkrone," Oeder wrote in a letter to von Haller a couple of weeks later.[9]

The presence of their excellencies did not do Oeder much good, though. In what resembles an early two-cultures manifestation, he was up against two opponents, only one of whom had any medical training. The other opponent's background was in theology. Oeder did officially complain about this but was merely told that in order to qualify for the medical doctorate, one must also master logic, metaphysics, and Latin.[10] Unlike more modern universities such as Oeder's alma mater, the University of Göttingen, Copenhagen still preferred dissertations to be written and defended in Latin. As Oeder was not very competent in Latin, it was comparatively easy to fail him. Later, some of his empirical results were published in French. Had the opponents at the oral defense paid more attention to the contents of Oeder's work rather than its form, he probably would have passed with excellence. But then, the emphasis on form may have been nothing but a bad excuse for getting rid of a foreign candidate in order instead to hire a local, Danish colleague.[11]

Oeder was clearly disappointed but not fazed by the treatment he received from the University of Copenhagen. Now, from the spring of 1752, in the pay of the royal household, he was asked to make plans for the introduction of botany as a new field of teaching and research. The obvious place to start, he informed his employers, was to build a good library in Copenhagen—one that would be open to the public but that would also serve as his own research library. "I do not yet have the necessary books and other aids to venture into publishing anything," he wrote to von Haller in May 1752.[12]

The lack of agreement among botanists concerning the naming of plants presented a major problem. Linnaeus's two important works, *Philosophia botanica* and *Species plantarum*, were published in 1751 and 1753, respectively, and it took another ten to twenty years for the Linnaean method of naming plants caught on. Until then, botanists had to consult the entire existing botanical literature to see whether a plant had already been named and described. Von Haller had taught Oeder to list the names of botanists (and the plants they had named) chronologically, the earliest botanists being mentioned first. According to von Haller, there was no reason to prefer the work of one botanist to that of another, and the names of plants and botanists should be listed only in relation to the books in which they were mentioned.[13] If the king wanted him to describe and register plants to be found in the Danish realm, Oeder therefore needed to have access to an up-to-date and well-stocked library.[14]

Important as the new library would be for him personally, Oeder also strongly promoted making it publicly accessible. The idea was that it would serve "the growth of the sciences and be of real and general use to the country."[15] As we saw, he was quite willing to share his library with the public (at least for a couple of days every week). Botany cannot "be of general use as long as it is a science only for the few," he insisted.[16] During his travels to Germany, Holland, France, and England to visit botanical gardens and to search for a good botanical artist for *Flora Danica*, he made sure to purchase books and journals that could be read by nonspecialists for the new library. A good number of these were in English and concerned the practical economics of plants, agriculture, and forestry, topics to which Oeder was to become increasingly attracted over the years.[17] When he traveled in England, he was impressed by how far the English had come in gardening and forestry—much further, he thought, than the Germans, for whom botany was still mostly a medical discipline; German botanical gardens primarily displayed medicinal herbs and plants.[18]

Educated Danes typically would know German and French but could not be counted on to know English, which Oeder had mastered. To accommodate a Danish readership, Oeder therefore made sure to buy many of these publications in a German or French translation. Of all the books he purchased in translation, by far the majority were in fact translated from English. This shows, perhaps more than anything else does, how much care Oeder took to make his library accessible to the lay, educated reader and "lover" of botany.[19]

Traveling in Norway

"Scientists have always had patrons with motivations of their own. . . . Utility— economic or otherwise—has long been a justification for the support of science, in terms of both finance and cultural approbation," writes Naomi Oreskes.[20] *Flora Danica* and the botanical library Oeder was hired to establish are good examples. In Chapter 1, we saw how crowdsourcing was used to gather plants across the Danish realm for *Flora Danica* and how sets of the finished work were sent to clergymen, grammar schools, and educated people in the hope that the work would find its way to members of the public. The motivation was utility—economic mostly, but mixed with a genuine wish to educate the public, many of whom did not have direct access to medical advice.

It was this double sense of utility that took Oeder to Norway after he had come back from his travels to Germany, Holland, France, and England. He left in May 1755 and more or less stayed in Norway until February 1760. His official orders were to gather plants as well as to explore geological and physical conditions, but already from December 1755, he started sending home to Copenhagen reports on socioeconomic issues as well.[21] One report concerns cases, in the Norwegian countryside, of pleurisy, tuberculosis, and smallpox. Many of those falling ill were young people living in faraway valleys without access to medical assistance. What they did have access to were the cures and remedies offered by quacks passing through. Danish authorities (Norway being at the time part of the Danish realm) ought therefore to intervene, Oeder advised.[22]

Such an intervention might be to support inoculation. Oeder described to his employers how a surgeon in Trondheim had successfully inoculated a number of smallpox patients between 1755 and 1756. It was not until 1796 that English country doctor Edward Jenner came up with a vaccine against smallpox, but others had been experimenting with finding a cure for a number of years. One of these was Oeder's former teacher, Albrecht von Haller, who had successfully inoculated his own daughter. Oeder was also aware that surgeons at Frederiks Hospital back home in Copenhagen (where Oeder himself would live for a while and establish the new public library) had been performing inoculation on a private basis.[23]

But it was England that was the prime mover. Already at the beginning of the eighteenth century, Lady Mary Montagu had sent dispatches home to England from Turkey, describing inoculation procedures performed there. She was staying for a couple of years in Turkey with her husband, who was the British ambassador to Constantinople (today Istanbul) at the time, and had seen with her own eyes how inoculation worked. Before going back to England, she had her three-year-old son inoculated. He became the first English person to undergo smallpox inoculation, and he never got the disease. Lady Mary had had smallpox as a young woman, and she had lost her favorite brother to the horrible illness, so she was determined to bring the technique of inoculation home with her.

Her contemporaries did not take much notice, though, Lady Mary's enthusiasm being "met with disdain by the English medical community. The reasons ranged from religious (what could Mohammedans teach Christians?) to medical (an untrained aristocrat lecturing physicians?) to economic (physicians of the day made a lot of money from useless smallpox treatments) to sexist (a female changing the thinking of men?)."[24] However, at the time of Oeder's reports from Norway, England was the leading country for smallpox inoculation, and young medical doctors from other countries would travel there to learn from English surgeons. Eventually, successful use of the smallpox vaccine would lead to a decrease in cases of smallpox, which was declared to be eradicated in 1977. "The history of smallpox holds a unique place in human health and medicine. One of the deadliest diseases known to humans, smallpox is also the only human disease to have been eradicated by vaccination."[25] It is interesting to note in this connection that the World Health Organization (WHO) today considers vaccine hesitancy to be one of the ten greatest threats to global health.[26]

Another intervention that Oeder suggested his Danish employers make was to increase the number of medical doctors in Norway. It was not just in the Norwegian valleys that there was a lack of medically educated personnel; the entire country counted only five officially employed health inspectors.[27] One possibility, Oeder wrote in a report from 1756, was to provide members of the clergy with some elementary medical knowledge. They—or at least their local constituencies—might be better off if instead of learning only theology, they were taught basic medical procedures such as how to bleed patients.[28]

A second possibility, which might in the end be more feasible, Oeder wrote in a later report, was to create a new kind of semiprofessional medical education, which would be more practice oriented and much less demanding than the ordinary medical education. The problem, as Oeder saw it, was that the latter was far too academic. Medical lectures took place in Latin, instead of in Danish or Norwegian, and such an education was so expensive that ordinary people could not afford it. Would it not instead be possible, he asked, to hold lectures in the local language and to create a much shorter, more practical, and less expensive education for young men of limited means? Knowledge of Latin would not be needed, and the trainees would be apprenticed to well-known and respected medical doctors from whom they could learn the trade "for the benefit of the common man."[29]

Looking toward Sweden, where almanacs describing common diseases and their treatment had been available to the public since 1740, Oeder also advanced the idea, after he had returned to Denmark, that the medical faculty of the University of Copenhagen organize prize paper competitions on popular medical themes and common illnesses. These would be open to medical doctors who would be encouraged to write in a nonacademic and straightforward way in Danish for nonspecialists. The prize papers could then be used as background reading for the apprentice doctors—and eventually be included in some of the most popular almanacs for general consumption.[30]

These constructive efforts by Oeder to remedy the lack of medical practice and knowledge in the Danish realm were never acted on by his superiors.[31] But, along with his hard work to create the botanical library and other botanical institutions in Copenhagen, they do show how serious he was about "the public benefit and use" of botany, plants, and medicine in general. Oeder would have given a nod of recognition to the vision and values statement of the Natural History Museum of Denmark, the contemporary successor to the botanical institutions he helped to establish, which wants "to empower citizens to connect with nature . . . to inspire, engage and enable people to enjoy, understand and care for the diversity of the natural world."[32] He would also have approved of the ongoing efforts of the museum to engage citizen scientists.[33] It was the Natural History Museum that ran the 2013 *Flora Danica* exhibition, mentioned in Chapter 2—an exhibition with a special focus on the interplay of science and art as one way of disseminating and engaging with scientific knowledge.

Article 15(2), Legally and Conceptually

Using art as a medium to convey scientific knowledge to a broader audience seems to be popular with museum visitors, as we shall see later. From a right-to-science perspective, the question is whether the integration of science and art can help break down some of the obstacles standing in the way of equitable access to and distribution and dissemination of science as a common good.

One obvious obstacle is the language scientists use to present new break-throughs and results to colleagues and other professionals via peer-reviewed publications and conference presentations and in the classroom. Often highly specialized and technical, such scientific lingo must be translated or curated in order for the public to understand it.

In 2003, the duty of researchers to disseminate their research and to take part in public debate was written into the Danish law on universities. The requirement adds to two other duties of Danish university researchers: research and teaching. The idea was and still is that Danish citizens ought to know what they get for their tax money. Since citizens pay, directly or indirectly, for the research done by scientific researchers and pay their salaries, the populace must be able to access the fruits of scientists' imagination and labor. Dissemination is no easy exercise, though. When a member of the Danish press asked Niels Bohr, a few years after he received the Nobel Prize in Physics, to explain briefly what the principle of complementarity was all about, Bohr reportedly answered, "I don't want to do that as it would just be short, clear, and wrong."[34]

Popular science writing is very different from writing research publications and the question is who should do it: researchers themselves, science journalists, popular science writers—or researchers in combination with journalists and popular science writers? Translation may be involved, moreover, typically from English, which is today the international scientific lingua franca, into other languages. Some progress has been made in translation efficiency by means of automated translation tools, but multilingualism in science is still inadequate.[35] It is essential, therefore, for translators to recognize the particular translation problems involved in the popularization of science and how these may affect the quality of translated works.[36]

Digital exclusion and lack of scientific literacy are additional obstacles to the dissemination of science as a common good. Today, public access to scientific information primarily involves digital media—making digital versions of new research publicly available by means of, for example, open access journals and repositories and mandatory open access policies that especially enable publicly funded research to be shared across the world. But digital divides in computer or cell phone use and in access to the Internet exist "for reasons of income, education, gender and geographic location."[37]

Even in areas of the world where people do have access to scientific knowledge on as well as off the Internet, scientific literacy is lacking. Many simply do not know how to tell real from fake science or pseudoscience: "Not only can people find it difficult to grasp new knowledge (largely due to the digital divide), but they also lack the necessary critical tools to question this knowledge (in terms of source and content) and assess its reliability. Therefore, both formal and informal educational institutions now have a central role to play in the scientific training of citizens and in teaching about the strict standards that govern how science is produced and disseminated in society."[38] Recognizing the importance of digital tools in disseminating knowledge and developing new methods for teaching scientific knowledge is clearly needed to help people become both digitally and scientifically literate. A number of scholars and educators are working on this issue.[39]

The ICESCR does not give any concrete suggestions for steps to be taken by states parties to achieve full realization of the right to science, but the 2020 General Comment No. 25 on science offers some guidance on what a supportive environment for "the conservation, development and diffusion of science" may entail.[40] What the CESCR terms "the right to participate in and enjoy the benefits of scientific progress and its applications" (that is, the right to science) contains both freedoms and entitlements and consists of five interrelated elements.[41] The first of these, availability, refers to the obligation of states parties "to take steps for the conservation, the development and the diffusion of science"—that is, to make certain that scientific progress actually takes place, and that the resulting scientific knowledge and products are widely disseminated.[42]

Accessibility, the second element, obliges states parties to eliminate barriers to participation in scientific progress and its applications. This includes furthering the access by marginalized populations to scientific education and providing information on the benefits as well as the risks that science and technology can generate.[43] Adequate resources should be directed toward making the third element, quality science—meaning "the most advanced, up-to-date and generally accepted and verifiable science available at the time, according to the standards generally accepted by the scientific community"— accessible for everybody, without discrimination.[44]

The fourth element concerns the acceptability of the scientific results. States parties are responsible for explaining results in a way that furthers their acceptance by various cultural and social groups, without compromising their quality. Acceptability furthermore relates to the tailoring of scientific education and scientific products to groups with special needs as well as the incorporation of "ethical standards in order to ensure [the] integrity [of scientific research] and the respect of human dignity, such as the standards proposed in the Universal Declaration of Bioethics and Human Rights."[45] Finally, the freedom that is indispensable for scientific research and creative activity, laid out specifically in ICESCR Article 15(3), makes up the fifth element.[46]

It is interesting that the General Comment mentions scientific freedom as one of the things that states parties must honor as a part of their obligation toward developing science. The issue of scientific freedom sometimes gets short shrift in discussions concerning the right to science, most people today being primarily interested in citizen science and the role of citizens in defining science policy. For scholars who care deeply about scientific freedom, it is therefore reassuring to see it mentioned in the General Comment also in the context of Article 15(2). Chapter 4 discusses ICESCR Article 15(3).

It is furthermore noteworthy that the General Comment emphasizes the positive duty of states parties to advance science through investment in education, science, and technology and through "allocating appropriate resources in budgets."[47] Especially important in light of present austerity measures, cuts in public expenditures for research, and large-scale privatization of scientific research by companies and nonstate actors, this wording points toward realizing the right to science by funding research and development that is not purely market driven. From a human rights perspective, both science and technology should be considered public goods that serve neither the power of states nor private profit but, much like education and health care, must be made accessible to everyone.[48] Austerity measures and cuts in public budgets for basic science may, as we shall see in Chapter 4, also qualify as retrogressive measures that violate states parties' core right to science obligation to respect scientific freedom.

Finally, of special note in the present context is the fact that the General Comment highlights the importance of scientific evidence being subjected to public scrutiny and citizen participation. Political decision making should be based on the best available scientific evidence, but the public should be actively involved in science policy and the scientific research process.[49] Not only is this good for democratic debate, but it may help generate and communicate socially and culturally relevant research topics, just as it may further transparent research and alert us to dual use science and technology.[50] A further advantage is that the active involvement of the public may increase scientific literacy and lifelong learning—just as it may draw attention to the importance of multilingual science.[51]

Dissemination of Science and the Right to Education

Together, the conservation, development, and diffusion of science form a crucial part of the right to science—one that may help explain why this particular human right is often considered a prerequisite for the realization of many other human rights. Two of these are the rights to health and to development, but no less important is the connection with the right to education—at least if we choose to see the right to science as being about more than only the material benefits and utilitarian value of science: "Through education, culture, and science, human beings collaborate to realize values of beauty, creativity, the search for truth and realization of a better tomorrow," writes Lea Bishop Shaver. "The value of science then, is not purely instrumental. Yes, science and technology have significant utilitarian value. They can be deployed to solve social problems and improve our material situation. But there is also a value inherent in the process itself, as with the educational process."[52]

In its support for the acquisition of scientific knowledge by as many people as possible, for as long as possible, the right to science overlaps with the right to education—and with Development Goal No. 4: Ensure inclusive and equitable quality education and promote lifelong learning opportunities for all. According to Cécile Petitgand and colleagues, "In fact, lifelong learning constitutes one of the enabling factors for achieving SDG [sustainable development goal] 4 in the sense that continuous education can enable everyone who was unable to benefit from quality education in elementary or high school to acquire the key tools for finding innovative solutions to the problems they encounter in their lives."[53]

Lifelong learning has always been a priority for the European Union. Between 2007 and 2013, for example, the European Commission supported the Grundtvig Programme, named after Danish minister and educationist Nikolaj Frederik Severin Grundtvig (1783–1872). Part of the European Commission Lifelong Learning Programme, which also encompassed the Comenius Programme for schools, the Erasmus Programme for higher education, and the Leonardo da Vinci Programme for vocational education and training, the Grundtvig Programme was aimed at providing adults with ways to improve their knowledge and skills, keeping them mentally fit and potentially more employable.[54,55] Among Grundtvig's core ideas, which are to this day alive and well in Danish so-called folk high schools, was informal adult education based on individual choices and without any top-down defined learning outcomes, grades, or exams. The first folk high school opened in 1844 in

Rødding. A few years later, in 1865, Askov Folk High School was founded as an extension to develop a particular focus on the natural sciences.[56] Folk high schools have since been used by adults, and increasingly seniors, as a lifelong learning educational alternative.[57]

The drafters of the UDHR linked lifelong learning and the right to education to the economic, social, and cultural rights listed in Articles 23 through 27. Together, these rights, of which the right to science is one, were the basis for "the full development of the human personality."[58] The right to science was included from the very beginning, but the obligation on states to take steps to conserve, develop, and diffuse science in realizing this right was inserted only during the drafting of the ICESCR. UNESCO submitted two different proposals for the article on the right to science, the longer of which read as follows:

> The signatory States undertake to encourage the preservation, development and propagation of science and culture by every appropriate means: (a) by facilitating for all access to manifestations of national and international cultural life, such as books, publications and works of art, and also the enjoyment of the benefits resulting from scientific progress and its application; (b) by preserving and protecting the inheritance of books, works of art and other monuments and objects of historic, scientific and cultural interest; (c) by assuring liberty and security to scholars and artists in their work and seeing that they enjoy material conditions necessary for research and creation; (d) by guaranteeing the free cultural development of racial and linguistic minorities.[59]

Sadly, these proposals were never thoroughly discussed. Had they been, we may have had more guidance concerning, for example, the framing and definition of the nature of science itself and the relationship between IP and authors' rights.

Unlike UDHR Article 26, which outlines the right to education and in detail explains why it is important and what it should do, Article 27 does not say anything about the ideological or philosophical direction of science. Various propositions were brought forward during the drafting process, but they were voted down. The Soviet Union suggested, for example, that the following wording be added to the text: "The development of science must serve the interests of progress and democracy and the cause of international peace and co-operation."[60] Some delegations were sympathetic to the idea behind the Soviet proposal, but it was defeated because many perceived it to be an attempt to use science to further a particular political ideology.[61]

The origin of the second paragraph of Article 27, which outlines authors' rights, seems to be the text submitted by Chile on behalf of the Inter-American Juridical Committee in 1947. Entitled "Right to Share in Benefits of Science," it drew attention to possible tension between public rights to participate and share in the benefits of science and IP rights. It read in full:

> Every person has the right to share in the benefits accruing from the discoveries and inventions of science, under conditions which permit a fair return to the industry and skill of those responsible for the discovery or invention.
>
> The State has the duty to encourage the development of the arts and sciences, but it must see to it that the laws for the protection of trademarks, patents and copyrights are not used for the establishment of monopolies which might prevent all persons from sharing in the benefits of science. It is the duty of the State to protect the citizen against the use of scientific discoveries in a manner to create fear and unrest among the people.[62]

Substantive discussion followed this proposal, the major parts of which never made it into the final document. This is sad, as the proposal touched on two issues that have since caused much debate: the tension between public rights to participate in the benefits of science and IP rights and the protection against dual use science. During the drafting of ICESCR Article 15 in the 1950s, the question of the protection of authors' rights came up again. While there was no major disagreement on the provision concerning the enjoyment of the benefits of scientific progress, it was the authors' rights provision on which opinions differed sharply.[63]

Science Beyond the Classroom: Science Museums, Science Communication, and Citizen Science

Lifelong learning about science topics often takes place outside the classroom in informal environments such as science museums and libraries. With their long history of educating the public and increasing public awareness, museums have learned to understand their visitors and their needs and have used this knowledge to develop sophisticated best practices and educational and disseminating infrastructures from which academics may learn quite a bit.[64]

Museums have not always emphasized interactive aspects centered on audience and visitor engagement. In outlining their more recent history, museum scholars talk about a number of paradigm shifts.[65] The first happened in the 1960s and 1970s and was a reflection of the political and cultural developments associated with the revolutionary events of 1968. Confronting head-on the notion of museums as highbrow places in which the elite could polish their taste and manners, curators and scholars now began to stress the social and didactic dimensions of their work. Intent on reaching a wider audience, they saw their role as that of "translating" their professional knowledge into exhibitions that would educate and teach visitors—especially schoolchildren and high school youths. The potential for museums to support and encourage education acquired great significance as national governments came to appreciate the social, cultural, economic, and individual benefits that flow from lifelong learning.[66]

The top-down method and one-way didacticism implicit in much of what was going on in and around museums in the 1970s and 1980s would become a major reason for criticism by the early 1990s. At this point, rather than educating the public and viewing visitors as (passive) receivers of education, museum scholars began to see visitors as customers, eventually as partners. In line with what came to be known as the "experience economy,"[67] museums began to orchestrate memorable events for their customer-visitors to provide opportunities for individualized challenges as well as for a personalized construction of meaning.[68] Furthermore, museums were increasingly seen as instruments for government policies on social inclusion, cohesion, and access.[69] For this, digitization came in handy; the advent of the computer enabled people of all ages to gain the skills and knowledge they need to contribute to and share in the information and communication age of the twenty-first century.

Digitization led to another paradigm shift as a result of which museums have come to be seen as places of dialogue between museum employees and the general public.[70] Research and understanding are no longer provided by museum researchers for the benefit of the public, but instead happen as a result of an active, two-way communication. Notions of authorship, creativity, and collaboration have become part of everyday culture rather than remaining in the hands of authoritative institutions.[71] Scholars of science communication talk about a move from public *understanding* of science and technology to public *engagement* with its practices, meanings, and implications.[72] Some maintain that ignoring lay knowledge can have financial costs for and may damage trust in STEM (science, technology, engineering, and medicine) museums, whereas others argue that incorporating public perspectives does not fundamentally challenge outmoded models of "science" and "society."[73]

The vote is still out on where the story of a move from public understanding of science and technology to public engagement will lead. Yet, what seems reasonably clear is that museums and academia share a number of things. Most notably, they both focus on education and research, including how to transfer knowledge and redesign academic research outcomes into accessible mediums—material, digital, and social—for the public. A key principle in the practice of knowledge dissemination is understanding your audience and their needs so as to be able to disseminate information accordingly. Museums have been very good at this, and scholars and scientists can learn from the audience and visitor engagement that informs the curatorial, communications, and public programming methodologies of museums.[74]

In a 2019 article, Clifton-Ross and coauthors explore what they call "research curation."[75] Comparing the dissemination of research outcomes to the communication of a curated body of work, this is a practice that "integrates contemporary curatorial and communications methods developed in museums with Internet Communication Technology best practices to strategically disseminate curated research outcomes to diverse audiences not typically reached through standard academic communication channels."[76] While the practice of curating research outcomes is not the same as museum curation, it makes use of information dissemination practices and public-facing aspects that have been shown to work well in museums, enhancing engagement and understanding.[77]

Museum visitors love the chance to speak directly with scientists, and for scientists, who do most of their research and teaching in a university setting, it is exciting to share their research directly with the general public.[78] When they design exhibitions, museum scientists have to be team players, recognizing other peoples' expertise and knowledge. They work with nonscientists more than other kinds of researchers typically do.[79] Artists make up one such group of nonscientists, and the integration of artwork in scientific exhibitions is an innovative form of science communication and diffusion. "Science describes the world in abstract terms, while the world is concrete, and only art, with its third language, is capable of bringing these two realities together," wrote Norwegian author Karl Ove Knausgård in a 2020 review of German artist Anselm Kiefer's work.[80] The dialogue between science and art, to which Knausgård is alluding, is growing.

At the Medical Museion, a combined museum and research unit at the University of Copenhagen with a focus on medical history and museology, a permanent exhibition, *Mind the Gut*, uses a blend of science, art, and history to highlight the complex relationship between brain, gut feelings, identity, and bowels, for example.[81] Human beings have always been fascinated by this relationship, and the exhibition shows how it has come to play a major role in contemporary science and in societal debates concerning health, treatment, and lifestyle choices.[82]

In connection with the mind and gut theme, the Medical Museion conducted a 2019 writing workshop, "Writing from the Gut," that explored the importance of "gut thinking" for creative writing, along with a meditation event, "Meditate with Microbes."[83] Involving audiovisual sculptures, the latter was a coproduction of artists and scientists that invited people to think about the relationship between how they perceive and interpret invisible micro-organisms and the factual existence of these organisms, in their bodies and elsewhere in nature.[84] Asking people to consider the relation of science to their own personal lives, these dissemination activities are fascinating ways to combine scientific learning, creativity, and ethical thinking.

Reflecting on this practice, professor and director of the Museion Ken Arnold and his colleagues are very pleased with the outcome. Yet, they do not want the Museion to become simply "a sci-art venue" as "the danger of adopting an overly dominant science-meets-art modality is that other disciplines and perspectives could be side-lined, for example leaving little room for enlightenment gained from the history or sociology or ethnography of medicine."[85] Approaches in addition to sci-art through which science could lead to palpable differences in people's lives should also be explored.[86] One such possible approach is citizen science.

Citizen Science

Citizen science is viewed by many as a highly promising way to reach new audiences and target groups. Citizen science "has amazing potential as an innovative approach to data gathering and experimental design, as well as an educational and outreach tool," conclude the authors of a 2019 opinion piece, for example.[87]

Citizen science encompasses a wide range of activities and projects carried out by universities, nongovernmental organizations, and individual citizen scientists, and professional citizen science networks have been established around the world (e.g., the U.S.-based Citizen Science Association, the European Citizen Science Association, and the Australian Citizen Science Association). Yet there is no official definition of the term—different people seem to mean different things when they refer to citizen science.[88] Lack of a standard meaning can lead to uncertainty when, for example, IP issues and scientific integrity are involved. Privacy issues and the abuse of personal data constitute further concerns, just as people have wondered whether the involvement of voluntary work on a private basis may be just another way for public institutions to save costs.[89] And what about quality control and the integrity of the scientific process? With citizen science, established rules of the scientific game that used to serve as a check on the quality of the research done may be upended.

Engaging individuals who have no formal scientific training, citizen science is today used in a variety of research projects around the world that collect, categorize, transcribe, and analyze scientific data, often with the use of digital media. Citizen scientists thereby obtain or manage "scientific information at scales or resolutions unattainable by individual researchers or research teams, whether enrolling thousands of individuals collecting data across several continents, enlisting small armies of participants in categorizing vast quantities of online data, or organizing small groups of volunteers to tackle local problems."[90] The advantages to the advancement of research and its applications are obvious: the involvement of citizen scientists in many parts of the world both reduces the costs and increases the speed of data collection, just as it accelerates the processes of analyzing the data collected.[91]

When it comes to what citizen scientists and society in general get out of it, the consensus is that citizen science can complement existing education frameworks by providing civic engagement, scientific literacy, and lifelong learning. As Rick Bonney and colleagues note, "Citizen science can provide opportunities for people of many backgrounds and cultures to use science to address community-driven questions."[92] For citizens, the result is an empowerment that helps demystify scientific activity and opens up new possibilities in life.[93] The prominent roles that science and technology play today makes scientific literacy even more important—for the individual, but also for society. As the drafters of the UDHR knew, an educated citizenry is a decided benefit—the best defense, in fact, against hatred and bigotry and of democracy.[94] Only a scientifically enlightened citizenry can make full and positive use, moreover, of the democratic right to be consulted on science policy issues, provided by Article 15(2) and the right to science in general.

For researchers, some worries remain. In addition to those already mentioned, concerns relate to the possible interference by citizen scientists and other nonscientists, as a part of their democratic mandate, in the free choice of research topics and the carrying out of this choice. Scientific freedom is one of the most important values to scientists. It is protected in Article 15(3)—and, as we saw, in that part of Article 15(2) that concerns the commitment to the development of science and technology for human benefit—and may sometimes clash with other parts of Article 15. When this is the case, a fair and equitable balance must be found. There will obviously be no scientific results and products from which the public may benefit unless scientist are granted freedom to follow their scientific theories and ideas. Against this must be weighed the very real possibility of dual use. This will be discussed in the following chapter.

Conclusion

From the beginning, science museums and libraries have been both local and global. This is especially true for institutions such as those started by Christian Georg Oeder in Copenhagen toward the end of the eighteenth century, which eventually developed into national institutions. European national museums, whether science, history, or art museums, have been at the center of nation-making and nation-building processes, their collections and displays reflecting and articulating dominant national norms and myths.[95] At the same time, they have aimed to be for everyone, for the whole world. Sir Hans Sloane, the donor of the British Museum's founding collection, wanted the museum to be a space for learning about and publicly debating the world and one's place in it.[96]

To many postcolonial scholars, the claims of the Hans Sloanes of this (museum) world are not to be taken seriously; they should be seen as purely pretentious and as being driven by an imperialist agenda to own, conquer, and exploit. There may be a good deal of truth in this, as we have seen with international controversies over repatriation and reparation issues, the underlying question invariably being who owns culture and cultural and natural heritage. Yet, we live in a global world with global problems that can only be solved globally. Human rights are universal, their language—like that of science—a global one. In the world of museums, ideas of the global museum have been debated for some time. The context is different today, however. Updated for the twenty-first century, the global museum wants to be inclusive, to tell local stories that are embedded in a global setting, and to reshape knowledge systems for public dissemination in line with multicultural and intercultural states and communities.[97]

The Museum for the United Nations–UN Live is one such new and visionary global museum. Believing that "to be truly global, we must be local everywhere," it hopes to put people everywhere in touch with the work and values of the United Nations, to help achieve its goals.[98] An independent not-for-profit NGO registered in Copenhagen, UN Live originated in 2014 and was formally endorsed by UN Secretary-General Ban Ki-moon in 2016. Self-described as "close to, but not part of, the United Nations," it is privately funded but also receives support from the Danish government and the city of Copenhagen.[99] For its pilot campaign, *My Mark: My City*, UN Live is working with key UN agencies as well as with locally placed innovators to come up with transformative solutions to the climate crisis.[100] UN Live is an interesting mixture of a global and local, publicly and privately funded, digital and analogue institution. Until an experience center and headquarters is built in Copenhagen, it exists and works as a digital platform that is free and accessible to everyone.

My Mark: My City (or at least the title) is presented in five languages on the UN Live website. Empowering learners at both global and local levels obviously requires more than one language. Multilingualism is an important issue in the non-English-speaking part of the world—not least when it comes to educating the public about scientific processes and the nature of scientific inquiry. Science is "today's currency for empowerment and social change," and "scientific understanding involves discussion, argument, reflection, and synthesis—challenging norms and reflecting on bias."[101] From a right to science perspective, linguistic exclusion should therefore be seen as adding to, or being a negative extension of, digital exclusion and the lack of scientific literacy.

Linguistic exclusion can also be a problem for scientists whose native language is not English. In a publish or perish era, it is increasingly important to get one's scientific articles published in the right international journals. But this can be difficult if one's field is, to take an example from my own scientific backyard, Danish history. As a Danish colleague in the history department once said to me, "who will write and care about Danish history if not Danish historians?" And whereas our fellow Danes have an obvious interest in publications in Danish concerning their own history, this is most probably not the case for people in other parts of the world. To this must be added the fact that not all scholars master English at a sufficiently proficient level to publish in English-language journals. Linguistic diversity may suffer as a result. In the Danish case, there is a justified worry that Danish will cease to exist as an academic language.

Toward the end of Chapter 2, I mentioned a publication by the Royal Danish Society of Arts and Letters.[102] With the exception of Georg Christian Oeder and a couple of his successors, all editors of *Flora Danica* were elected members of the society.[103] When it was founded in 1742, the issue of language came up. A few of the members of the new society argued in favor of the publication of its proceedings in Latin, the learned, academic language of the day. But for the majority of their colleagues, the most pressing issue was the duty toward their fellow Danes, who must be informed about what was going on in the world of science. As professor of mathematics Christen Hee expressed it in 1759, the publications of the society should "aim to further scientific knowledge within the Danish nation, not to boast to foreigners, to which end the Society should never publish anything in Latin."[104]

The Hee quotation is taken from an essay by Helge Kragh, "Between Provincialism and Internationalism: The Natural Sciences in Denmark During the Period of Enlightenment,"[105] a title that aptly describes the state of Danish scholarship in Hee's time. But it also captures the dilemma posed to this day to academics whose first language is not English: whereas the language of science and scholarly discussions is global (English), the dissemination of scientific results to the public takes place at the local level. In order to honor the obligations outlined in ICESCR Article 15(2), each individual scientist must be able to master two very different versions of science talk.

In this chapter, I have looked at the implications of Article 15(2) for the public. Next, I turn to the perspective of the individual scientist and to the issue of scientific freedom.

CHAPTER 4

===

Scientific Freedom

The States Parties to the present Covenant undertake to
respect the freedom indispensable for scientific research
and creative activity.
—Article 15(3) International Covenant on Economic,
Social and Cultural Rights

In this chapter, I explore whether it is possible to find a fair and equitable balance between the rights and needs of the public and those of scholars and scientists.

Experience shows that it is often curiosity-driven research that later becomes the basis for solving real-world problems. Basic, nonstrategic or nontranslational science, which comes with no obvious strings attached and no obvious demand for immediate utility, has proven crucial for humanity. But the problem is that there is no clear way to tell which basic scientific research will lead to major breakthroughs. Many of the key advancements of modern science are based on knowledge gained in basic science by people who were uncertain of where their research was leading and who worked with no clear time frame.

For example, the development of computer chips would have been impossible without quantum mechanics, just as today's GPS devices build on Einstein's theory of relativity. A third example is X-ray technology, the use of electromagnetic radiation, which was discovered in 1895 by Wilhelm Röntgen and is today widely used in medicine, materials analysis, and devices such as airport security scanners. Examples from medicine include penicillin, accidentally discovered by Alexander Fleming in 1928, which, though not widely distributed until after World War II, began the era of antibiotics, and the smallpox vaccine that was introduced by Edward Jenner in 1796. The first successful vaccine to be developed, it is today widely regarded as the foundation of immunology.[1]

For all the scientists behind these advancements, scientific freedom was crucial. It still is. For most scholars today, scientific freedom is the starting point for everything they do, and they worry about the increasing preference by funding bodies for applied research that can promise practical outcomes and products of immediate value. "The freedom indispensable for scientific research and creative activity" is codified in ICESCR Article 15(3).[2] States parties to the covenant must respect this freedom. In General Comment No. 17 (2005), the Committee on Economic, Social and Cultural Rights (CESCR), the monitoring body of the ICESCR, referred to a possible future general comment on Article 15(3).[3] The idea of writing an independent comment on the importance of respecting artistic and scientific freedom has not been followed up. Instead, the CESCR decided to treat all four parts of Article 15 in its General Comment No. 25 on Science (2020). From a right-to-science perspective, one important question is whether states parties and others promoting strategic research funding over—and at the cost of—funding for basic research may be said to be in violation of Article 15(3).

Sometimes, the preference for strategic research funding is explained or legitimized with reference to pressing societal needs. The best example is the urgency for the development of atomic weapons to end the Second World War and the scientific and engineering talent involved. A more recent example is the decision by the EU to integrate, or mainstream, climate change mitigation and adaptation into all major EU spending programs to help achieve its climate goals. This includes programs under research and innovation.[4] As citizens of both their own local communities and the world, scholars appreciate the value of such strategic research—just as they understand the democratic need for citizen science and citizen impact on science policy, as discussed in Chapter 3. Strictly speaking, though, both strategic climate research and citizen science impose limits on basic research and on the human right to scientific freedom.

Of basic importance to any discussion of scientific freedom are the definitions of "science" and "scientist." The first two parts of this chapter look into definitions that are relevant to the human rights context. The first part is intended as a prelude to the second part, in which the issue of definitions will be approached more concretely. First, I look at the disagreement, already alluded to in previous chapters, between Georg Christian Oeder, first editor of *Flora Danica*, and Carl Linnaeus concerning plant taxonomy. The practice of identifying, classifying, and describing plants has played an important role in botany, transforming it into an independent scientific discipline, distinct from medicine, agriculture, and traditional herb and plant lore. The debate between Oeder and Linnaeus remains of historical value today and may help us approach the discussion of what is and what is not considered science. The relationship between traditional knowledge and science, going on in disciplines such as ethnobotany and other qualitative research areas sensitive to cultural issues, for example, is of great current concern.[5] Local or traditional knowledge comes into play in many different ways around the world as we try to find solutions to climate change and other environmental catastrophes such as COVID-19.

The second part of the chapter focuses on the definitions of "science" and "scientist"—as well as "scientific freedom"—offered in relevant international human rights documents. Most prominent among these are CESCR General Comment No. 25 on Science and the 2017 UNESCO Recommendation on Science and Scientific Researchers, as well as the 2012 special rapporteur's report on the right to science, as we have seen in previous chapters. Regional human rights instruments such as the EU Charter of Fundamental Rights are also pertinent. The Freedom of the Arts and Sciences, outlined in Article 13 of the EU Charter, mentions respect for both academic and scientific freedom. This raises the question of the relationship between these two freedoms. Although they are often used interchangeably in public debate, they are not the same, and they have different historical roots.

In all of these human rights instruments, necessary restrictions on scientific freedom feature prominently. In order to safeguard basic human rights principles such as human dignity and nondiscrimination, prior informed consent, confidentiality of data, and other kinds of protection from dual use research are needed. The third and last part of this chapter concerns the danger of dual use and the responsibility of researchers to avoid unnecessary risks to people's health and well-being. From a human rights perspective, as I have already argued in previous chapters, scientific responsibility is an alternate aspect of scientific freedom. I will discuss dual use in both a narrow and a more general way. The original meaning of dual use referred to technologies that can have both military and civilian uses. Today, many people tend to think about dual use science and technology from a broader, more ethical perspective. They use it to distinguish between beneficent and legitimate or valuable uses, on the one hand, and malevolent and destructive uses on the other.

When it comes to dual use, the interests of the public sometimes clash
with those of scientists. Whereas scientists cherish their scientific freedom,
the public calls for important restrictions of that very freedom to avoid dual
use or to secure human dignity in experiments. On this issue, as on the topics
of strategic funding and citizen science, we must acknowledge the clashing
human rights interests involved in order to find equitable and just solutions.

Plant Taxonomy, Disagreements Between Oeder and Linnaeus, and the Relationships Between Traditional Knowledge and Science

"The beautiful plants were always put first in the *Flora Danica* fascicles,"
writes Henning Knudsen. "Then, as now, it was important to catch the attention of the reader, and this was done through the immediate appeal of the
beautiful plants. After the inviting plants came the grasses, and then the small
and much less known algae, mosses, lichens and fungi, which were put at the
end of the fascicles."[6]

The invitation to subscribe to *Flora Danica*, a four-page folder printed in
Danish and French that Oeder sent out in 1761, included the first plate in the
work, an illustration of the cloudberry plant. Related to the raspberry plant,
which also grows wild in Denmark, cloudberry can be found only in a few
places in the northern part of the country. It is much more common in Norway (which was part of the Danish realm in 1761) and other mountainous
northern regions, where it cannot be cultivated but is picked berry by berry
and used for jam, desserts, and a liqueur.[7] With a high vitamin C content,
cloudberries played an important role for public health before citrus fruits
became common. Its flavor is unique, and cloudberries are today an important ingredient in New Nordic cuisine.[8] Beautiful, wild growing, healthful, and
popular, the cloudberry plant was indeed a good choice on Oeder's part.

If, in order to sell more subscriptions, Oeder allowed nonscientific factors
such as beauty to influence the order of appearance of the plates making up
each fascicle of *Flora Danica*, he was much stricter when it came to naming
and describing the plants on those plates. Outlining the various botanical
systems of classification in use in his time in the first of the five planned parts
of the *Flora Danica* project, the introduction to the science of botany that was
published in Danish, Latin, and French in 1761 (see Chapter 1), Oeder con-
trasted his own system with that of Linnaeus. Whereas his Swedish colleague

used artificial classifications, argued Oeder, his own classification was based on empirical observation and mirrored what he saw as the true, natural relationship between plants.[9]

One of the people who influenced Oeder was the French scientist Georges-Louis Leclerc, Comte de Buffon, whom he met when he traveled in France to find illustrators for *Flora Danica* and books for the new public library in Copenhagen. Buffon, who was famous in his own day but is now mostly known as a precursor to colleagues such as Darwin, argued for a biological species concept and attacked what he considered to be essentialist and artificial concepts.[10] Another influence on Oeder was his teacher and mentor, Albrecht von Haller, in Göttingen, whose views on taxonomy were similar to those of Buffon. Von Haller emphasized even more strongly than Buffon did the variability of species as an aspect of taxonomy—something, he claimed, to which Linnaeus did not pay sufficient attention.[11] We know from an exchange of letters between them that von Haller found Linnaeus's binary nomenclature and sexual system arbitrary, rigid, and not rooted in nature. According to Mathias Persson,

> Aside from the personality-related factors, the conflict was informed by deepseated differences in the understanding of botany and, ultimately, the world. Firstly, in contrast to Linnaeus, Haller did not conceive of any consistent hierarchy between the various levels—class, order, species—within his system. Secondly, Haller's system was "natural"; he strove to reveal the multitude of connections among plants and to pay attention to as many of their features as possible, not just a few characteristics selected a priori, like the sexual organs. Thirdly, Haller refused to abandon the traditional, polynomial nomenclature and adopt Linnaeus' binary names, an approach which over time rendered him isolated and obsolete in botanical circles. The civil war in the republic of botany thus emanated from a variety of partly overlapping conditions.[12]

As he had learned from his teacher, Oeder cited all names used for each plant in chronological order, with the binary Linnaean name last (see Chapter 3). This earned him the scorn of his Swedish fellow botanist, and angry letters were exchanged between Oeder and Linnaeus with regard to the issue of nomenclature.[13] In some parts of the *Flora Danica* project, however, Oeder did use Linnaean binary nomenclature. This leaves the impression, per Ib

Friis, that "Oeder was aware of the *practical* use of Linnaean binary nomen-clature" (emphasis added).[14] So were his successors as editor of *Flora Danica*. Yet, even though virtually all botanists were using Linnaean nomenclature by the end of the eighteenth century, and even though they applied this nomen-clature in their other scientific work, Otto Friedrich Müller (editor from 1775 to 1782) and Martin Vahl (editor from 1787 to 1799) both followed Oeder's way of listing names chronologically when they worked with *Flora Danica*. So did Jens Wilken Hornemann (editor from 1810 to 1840)—at least in his first few fascicles. From 1810, he used only Linnaean nomenclature, however, and after 1840 all subsequent editors applied the Linnaean system.[15]

The disagreement between Linnaeus and Oeder, Buffon, and von Haller concerning taxonomy has never been completely resolved. It was Linnaeus's system, not that of his critics, that became the standard.[16] However, the prob-lem with hierarchical levels (class, order, species) and the artificiality of the Linnaean system, as well as the variation within and hybridization between species, to which Oeder, following Buffon and von Haller, tried to draw atten-tion, has since been raised by other botanists. "Even today," writes Ib Friis, "we have not completely developed the Linnaean nomenclature to reflect the natural variation in plants; unlike Oeder, the rebel, we have resigned our-selves with that part of the Linnaean tradition."[17]

With the introduction a century after Linnaeus of the theory of evolution by natural selection, Charles Darwin and Alfred Russel Wallace fundamen-tally changed biology. Yet, whereas the conceptual ground was transformed in various ways, the distinctly preevolutionary Linnaean system remained much the same. As some botanists see it, the Linnaean system has been useful and flexible enough to allow the theory of evolution to be formulated, and it has provided a great deal of stability over the years. Others consider the Linnaean hierarchical naming conventions stemming from categorical ranks totally outdated and want to abandon them as "biologically meaningless arti-ficial constructs that impede scientific progress."[18]

The discussions that took place between Linnaeus and Oeder, Buffon, and von Haller were discussions among fellow scientists. They concerned scien-tific aspects of botany—who has the most scientifically correct approach and, more generally, what is or is not science—and coincided with efforts to turn botany into a serious, scientific subject. The first university chairs in botany were established in Europe during the eighteenth century, the idea being to separate botany from medicine and to turn it into an independent scientific discipline. Oeder would have become Denmark's first professor of botany had not the University of Copenhagen prevented it, as previously described.

Scientific dissemination also played a role. Linnaeus made botany accessible to generations of the public. Botanical communication needs a recognized nomenclature, especially for the units in most common use. Even if scientifically rigid and hierarchical, as his critics have claimed, Linnaeus's classifications of plants provided such a nomenclature, a pragmatic tool for people who might not otherwise have become drawn to botany.[19] It also gave scientists an instrument for disseminating scientific knowledge, and without a comparatively simple method for disseminating scientific results, there is no access for the public to scientific results.[20]

Touching on and involving both the "diffusion of science," mentioned in Article 15(2), and the various meanings of "science" and "scientist" that inevitably come up in debates concerning "the freedom indispensable for scientific research," outlined in ICESCR Article 15(3), the battles over taxonomy may thus be viewed as early right-to-science disagreements.

The Relationship Between Traditional Knowledge and Science

When it comes to the identification of plants and quality control, taxonomy and methods of authentication are as important today as they were in Oeder's and Linnaeus's time.

Plants are fundamental to human existence. Writes Kathy J. Willis, "Plants underpin all aspects of our everyday life—from the food that we eat, to the clothes that we wear, the materials we use, the air we breathe, the medicines we take and much more."[21] Since prehistoric times, medicinal plants and herbs have been used in traditional medicine practices, and in many rural communities around the world people are still relying on traditional knowledge and plant-based medicines for their primary healthcare because they are affordable and accessible. But herbal medicines are also becoming increasingly popular in many urban areas, both in the West and elsewhere.[22] This popularity is no doubt related to the mounting interest in ecology, biodiversity, and all things environmental.

International trade in herbal medicine is growing as a result. This raises a number of questions and issues. One of these concerns the need for quality control and for more research to evaluate the medicinal properties of the plants in question and their potential as new drugs.[23] For the authentication of herbs and plants and their ingredients, taxonomy, or the correct labeling of plants, is important. Writes Bob Allkin, "Product labelling is frequently misleading, with the trade name 'ginseng,' for example, referring to 15 different species of plant, each with its own particular chemistry and therapeutic properties. Substitution by a Belgian clinic of one Chinese medicinal herb ('Fang Ji') with another sharing the same name, led to over 100 patients requiring kidney dialysis for the remainder of their lives."[24] Illustrations accompanied by detailed descriptions may help authenticate plants and identify counterfeit plants.[25]

Another issue of concern with regard to the growing international trade in herbal medicine is bioprospecting—that is, looking for ways to commercialize biodiversity through, for example, exploring traditional and indigenous knowledge. If done well and as a cooperation between research groups, public and private organizations and companies, and local groups with complementary knowledge, bioprospecting can lead to biodiversity-driven "evolution and sustainable use of medicinal plant diversity for unmet medical needs."[26] In the past, traditional forms of creativity (traditional knowledge, genetic resources, and traditional cultural expressions) have not always been recognized and protected from misappropriation by commercial interests, however. I shall come back to the way in which bioprospecting, if not well managed, can lead to disregard for the rights, knowledge, and dignity of local communities in Chapter 5.

In order to deal respectfully with traditional forms of creativity and bioprospecting, bridge building between customary and scientific (western) knowledge is needed. This is where ethnobotany comes in. Although all cultures have investigated plants and their uses for thousands of years, as an academic discipline ethnobotany dates back only to the late nineteenth and early twentieth centuries. As Danish ethnobotanist Vagn J. Brøndegaard (1919–2014) defined it, ethnobotany comprises "the traditional names of plants, their use during yearly holidays and feasts as well as in medicine, in home industry, craftsmanship, children's games, customs, myths and superstition, names of places, heraldry, sayings, prose, and poetry."[27]

Brøndegaard was a self-taught and self-proclaimed ethnobotanist, whose work was recognized by professional scholars only late in his life. A prolific writer of both scientific and popular articles and books, he maintained that it is not only possible but also helpful to establish links between traditional and scientific knowledge formation and production. In a 1969 article, whose topic was plant-based contraceptives, he wrote, for example, that "With so-called primitive peoples as well as in European folk medicine is gathered a large amount of empirical knowledge, which, if used rationally, could benefit modern medical science. Modern medicine has already received many valuable impulses from folk medicine. . . . [However,] most medical doctors and pharmacists still think of this 'materia medica' as superstition or, at least, as not worthy of scientific research."[28] Brøndegaard is today considered a pioneer of Scandinavian ethnobotanical research, and his major work on botany, *Flora and Fauna: Danish Ethnobotany*, in four volumes published in 1979, especially has been an inspiration to many.[29] Like other ethnobotanists, he sought throughout his life to understand how people interact with the environment and obtain plant resources to meet not only their physical but also their cultural needs. One of his topics is children's games that include plants—games like "She loves me, she loves me not," referred to in Chapter 2, which involves the marguerite or oxeye daisy and has been played by children and young people down through the ages. As illustrations of the relationship between human beings and nature, Brøndegaard argued, these games have been overlooked or underestimated by scholars. Ordinary occurrences of everyday life are often the ones that can tell us the most about what matters and has been useful to people. Much like the knowledge about herbal medicine and practices gathered by indigenous and other local communities, children's games rely on experience that can be scientifically tested by western medicine, then be made useful to everyone.[30]

With its many references to and quotations from literature and poetry, *Flora and Fauna* is as much a literary or artistic piece of work as it is a work of natural history. Brøndegaard clearly paid no attention to the nature-culture dichotomy or the schism between objective and subjective knowledge. Work by ethnobotanists such as Brøndegaard and by "cultural botanists" may help reconcile the two-cultures divide, argues John C. Ryan.[31] Encouraging the exchange between the arts, humanities, and botany, cultural botanical research may not only help revise "the technicized" view of plants since Linnaeus; it may also lead to "a new contract between humanity and the Earth [that] would entail a shift in power structures such that 'the natural world will never again be our property, either private or common, but our symbiont.'"[32]

Ryan here echoes the arguments of a number of biologists and historians of science who have written on the age that has come to be known as the Anthropocene.[33] From a human rights perspective, where does that leave traditional knowledge?

Article 15(3), Legally and Conceptually

In the discussion paper drafted by CESCR and forming the point of departure for a day of general discussion in Geneva in October 2018, question 16 had the following wording: "Not all kinds of knowledge are science. How should we consider other kinds of knowledge or cultural traditions that are not science? What is the relationship between cultural rights and science? How should we consider the issue of pseudo sciences?"[34]

The discussion paper was intended as part of a consultative process to help CESCR draft a new general comment on ICESCR Article 15, outlining the measures to be adopted by states parties to ensure full compliance with this article. In the 2005 General Comment No. 17 on the Right of Everyone to Benefit from the Protection of the Moral and Material Interests Resulting from any Scientific, Literary or Artistic Production of Which He or She is the Author (Article 15(1)(c) of the ICESCR), CESCR explains that the RtS will be "partly explored in this comment and partly in separate general comments on *article 15, paragraphs 1 (a) and (b), and 3*, of the Covenant" (emphases added).[35] The general comment on Article 15(1)(a) came four years later, in 2009. The idea of writing a general comment on Article 15(3), which would specifically highlight the importance of respecting artistic and scientific freedom, has so far not been followed up. For now, though focusing on Article 15(1)(b), in the 2020 comment on science, the Committee seeks to address, if only in broad and general terms, all four parts of Article 15 as these relate to science.[36]

Like other general comments, the General Comment on Science details the CESCR opinion on the normative content, elements, and limitations of this right. It also contains a section on the core obligations that states parties have with regard to this right. One of these is the obligation to "establish protective measures in relation to messages from pseudoscience, which creates ignorance and false expectations among the most vulnerable parts of the population. . . . For instance, when parents decide not to vaccinate their children on grounds the scientific community considers false."[37] Unlike the discussion paper, the general comment separates the issue of pseudoscience from the topic of other kinds of knowledge or cultural traditions that are not considered to be scientific. These latter are addressed both directly and more indirectly in other parts of the general comment.

In the subsection "Traditional Knowledge and Indigenous Peoples," for example, the comment acknowledges the important role that local, traditional, and indigenous knowledge plays "in the global scientific dialogue . . .

especially regarding nature, species (flora, fauna, seeds) and their properties."[38] Of special importance for addressing the issue raised in the discussion paper (concerning how to consider other kinds of knowledge or cultural traditions that are not science) is paragraph 41, which says that "indigenous peoples and local communities all over the globe should participate in a global intercultural dialogue for scientific progress, as their inputs are precious." This should happen in such a way that the rights of indigenous peoples are respected and protected, including "their land, their identity, and the protection of the moral and material interests resulting from their knowledge of which they are authors, individually or collectively."[39] Though valuable in and of itself, in other words, indigenous and local knowledge is not science.

This becomes even clearer when we look at how the general comment describes science. Repeating the wording of the 2017 UNESCO Recommendation on Science and Scientific Researchers, the general comment defines science as "the enterprise whereby humankind, acting individually or in small or large groups, makes an organized attempt, by means of the objective study of observed phenomena and its validation through sharing of findings and data and through peer review, to discover and master the chain of causalities, relations or interactions; brings together in a coordinated form subsystems of knowledge by means of systematic reflection and conceptualization; and thereby furnishes itself with the opportunity of using, to its own advantage, understanding of the processes and phenomena occurring in nature and society."[40]

In the 2017 UNESCO recommendation, this definition of "science" is followed by a definition of "the sciences" as a "complex of knowledge, fact and hypothesis" that can be validated, either immediately or later, and that includes the social sciences in addition to the natural and health sciences.[41] The general comment retains this definition. It also echoes the formal mention of "the human right to science" in the 2017 UNESCO recommendation as well as its acceptance that science should be considered, and has value as, a public good. Encompassing both natural and social sciences, science refers to the process and methodology of "doing science" as much as to the results that follow, according to the general comment. This means that "Though other forms of knowledge may claim protection and promotion as a cultural right, knowledge should only be considered as science if it is based on critical inquiry and open to falsifiability and testability. *Knowledge which is solely based on tradition, revelation or authority, without the possible contrast with reason and experience, or which is immune to any falsifiability or intersubjective verification, cannot be considered science*"[42] emphasis added).

Those areas whose knowledge production is not empirical—not "testable and refutable," as the 2012 special rapporteur report on the right to science puts it[43]—do not fall within the definition of science provided in the relevant human rights instruments. This includes much of the arts and the humanities as well as those areas of the social sciences that excel in aesthetic and qualitative rather than quantitative research. It furthermore includes traditional knowledge, as we saw. This omission may provoke scholars for whom science ought to encompass all scholarly activities, including those within the arts and the humanities.

From a human rights perspective, these scholars and their activities are instead offered protection by the right to culture as well as by an idea of academic freedom that can be traced back at least to the early Enlightenment period, especially to discussions at German universities. The German word for science, *Wissenschaft*, refers to all scholarly activities across the board, including those taking place in the arts and humanities. I come back to the difference between the German *Wissenschaft* and the Anglo-American *science* in the next section.

According to all three recent human rights instruments, which focus directly on the right to science—the 2020 General Comment, the 2017 UNESCO recommendation, and the 2012 special rapporteur report—a number of other human rights are relevant for our interpretation of this particular human right. First, and most obviously, the other two rights mentioned in ICESCR Article 15(1), the right to culture and authors' rights, are of immediate significance for the right to science. In the opinion of Special Rapporteur Farida Shaheed, as we saw in Chapter 2, "the rights to science and culture should be read together."[44] Cultural rights are about furthering human knowledge and creativity, as well as about enjoying the fruits of the creativity of others. This is where authors' rights come in and where the three parts of Article 15(1) are linked.[45]

Because of the enormous impact of science and technology in our daily lives, the right to science is essential for the realization of various other rights as well. Among these are the rights to health, water, housing, and education, as well as the right to development and the emerging right to a clean and healthy environment.[46] In addition, and again reflecting the importance of science and technology today, the right to science is connected, per Shaheed, with "freedom of expression, including the freedom to seek, receive and impart information and ideas of all kinds, regardless of frontiers, the right of everyone to take part in the conduct of public affairs, directly or through freely chosen representatives, and the right of all peoples to self-determination."[47] Especially, freedom of expression has traditionally been linked to both academic and scientific freedom.

Academic Freedom and Its Relation to Scientific Freedom

Scientific freedom has historically been closely related to the idea of academic freedom, which can be traced, as mentioned, to the early Enlightenment period.[48] Throughout the eighteenth century, the concept of academic freedom was widely discussed, especially at German universities such as the Universities of Halle, Jena, and Göttingen, Oeder's alma mater, founded in 1734. These discussions formed the background for the development, beginning in the early nineteenth century, of the German research university. The German model involved several innovations suggested by Wilhelm von Humboldt, who based these on ideas derived from the philosophy of Friedrich Schleiermacher. These include the equal standing of students and professors in seminars, so that they can work together (just as graduate students and their supervisors commonly do today), the freedom of professors to teach as they see fit, and the corresponding freedom of students to learn what and as they please.

The German research universities were considered to be among the finest in the world and therefore appealed to many foreign students. American students, who were used to collegiate universities, especially were attracted to the German model. When they returned home, they took with them the concepts of academic and scientific freedom—only to find that some of the basic ideas behind the German research university model were not easily transplantable across the Atlantic. To start with, the German word *Wissenschaft*, translated into English as *science*, had a broader meaning in that it included the social sciences as well as the humanities and, to some extent, the arts. This is still the case in modern German, as it is in other Germanic languages such as my own Danish.

The German understanding of academic freedom consisted of three key concepts: *Lehrfreiheit* (freedom of teaching), *Lernfreiheit* (freedom of learning), and *Freiheit der Wissenschaft* (scientific freedom). *Lehrfreiheit* referred to the legal status of associate and full professors who, as civil servants in the German system, were allowed to operate outside the laws and regulations affecting other civil servants. In the United States, however, professors were employees of the board of governors of the university. *Lernfreiheit* was understood to mean that the universities had no authority over students beyond that of administering exams and granting degrees. Here, too, the American system differed in that colleges typically retained *in loco parentis* responsibilities over their charges. Finally, implicit in the idea of *Freiheit der Wissenschaft* was academic self-government: the independence of the university as an institution from interference by political or other outside powers. In the United States, the university was dependent on the whims of the members of the governing body, who were typically donors or political actors. These national differences meant that German research university models and concepts such as academic freedom had to be adapted to a different American context.

Protecting in Article 5 the freedom of "the arts and sciences, research and teaching," the German constitution of 1949 reflects this historical legacy.[49] So does—perhaps—Article 13 of the European Charter of Fundamental Rights (2000). Whereas UDHR Article 27 does not create a right to scientific freedom as such, and ICESCR Article 15(3) mentions scientific but not academic freedom, Article 13 of the European Charter talks about both scientific and academic freedom: "The arts and scientific research shall be free of constraint. Academic freedom shall be respected." These freedoms are not defined, and they are not to be found in the European Convention on Human Rights (ECHR) from 1950. The Explanations to the Charter confirm, however, that they are deduced primarily from the rights to freedom of thought and expression, set out in Articles 9 and 10 of the ECHR, and must be subject to the limitations authorized by these articles.[50] In a famous decision in 1976, the European Court of Human Rights ruled that freedom of expression constitutes "one of the essential foundations of [a democratic] society, one of the basic conditions for its progress and for the development of every man."[51]

Drawing, perhaps, on a conception of scientific freedom based on the German *Freiheit der Wissenschaft* model, Klaus Beiter argues that academic freedom is a special part of scientific freedom that applies only to academics who teach or do research at universities.[52] Freedom of science applies both within and beyond universities to all scientific endeavors. Whether a person enjoys academic or scientific freedom therefore has nothing to do with what field of study they engage in or what kind of approach they take. Instead, what matters is this: "Are we dealing with science (and this includes the humanities and social sciences) within or beyond the university? In the former case, we are dealing with academic freedom (as a concretised freedom of science); in the latter case, with 'ordinary' freedom of science. 'University' here does not solely [denote] 'traditional' universities, but all higher education institutions broadly dedicated to the triad of learning, teaching and research as facilitating *notably disinterested science*"[53] (emphasis added).

In the UN system as in the European system, freedom of expression is viewed as a key democratic right that protects academic freedom. In addition to being part of the guarantee of free speech under ICCPR Article 19, academic freedom is considered to be central to the right to education.[54] CESCR published a general comment in 1999 on the right to education, which affirms that this right can be enjoyed only if it is complemented by the academic freedom of staff and students.[55] This makes good sense as education and academic freedom mutually reinforce each other and are directed toward "achieving the 'greater' overarching goals of science—science, rather than 'instruction' being what universities are for."[56] Another authoritative source on academic freedom is the 1997 UNESCO Recommendation Concerning the Status of Higher Education Teaching Personnel, which, though not binding, reflects something close to an international consensus on the meaning of academic freedom.[57] More recently, the special rapporteur on the promotion and protection of the right to freedom of opinion and expression published a report on the freedom of opinion and expression aspects of academic freedom.[58]

With regard, more specifically, to scientific freedom, the 2020 General Comment on Science, relying once again on the 2017 UNESCO Recommendation on Science and Scientific Researchers, outlines a number of dimensions of Article 15(3). Overlapping to a certain extent with academic freedom, as defined in the 1997 UNESCO recommendation, the freedom indispensable for scientific research involves "protection of researchers from undue influence on their independent judgment; their possibility to set up autonomous research institutions-and to define the aims and objectives of the research and the methods to be adopted; the freedom of researchers to freely and openly question the ethical value of certain projects and the right to withdraw from those projects if their conscience so dictates; the freedom of researchers to cooperate with other researchers, both nationally and internationally; the sharing of scientific data and analysis with policymakers, and with the public, wherever possible."[59]

Scientific freedom is not unlimited, however. The UN system has worked persistently toward protecting people from the negative effects of dual use science. The 1997 UN Declaration on the Human Genome and Human Rights prohibits human cloning, for example, just as the 2005 UNESCO Universal Declaration on Bioethics and Human Rights aims to make sure that biomedical activities comply with international human rights law.[60]

Possibility of Limitations: Dual Use, the Precautionary
Principle, and Retrogressive Measures

As already mentioned, the concept of dual use was first applied to describe rocket science during the Cold War: the technology that can put an astronaut into space is the same technology that launches intercontinental ballistic missiles. Dual use since then has referred broadly to any technology that can be used in the pursuit of two separate aims, but often it is used in a more narrow sense to describe technology with a civilian and a military use. The Internet, GPS systems, and many other modern civilian technologies are examples of military technology that has been modified and used for civilian purposes. The debate over treatment and enhancement in genetics and synthetic biology stems from the fact that advances in these fields might be used to improve biological traits in healthy, just as much as in sickly, individuals.

The fact that some technologies and products of science can be put to both benign and deleterious uses is recognized not only in the 2012 special rapporteur report on the right to science but also in the 2017 UNESCO recommendation and the 2020 General Comment on Science. The latter opens this way: "The intense and rapid development of science and technology has had many benefits for the enjoyment of economic, social and cultural rights (Hereinafter: ESCRs). At the same time, the risks and the unequal distribution of these benefits and risks, have prompted a rich and growing discussion on the relationship between science and ESCRs."[61] In view of this, certain limits are necessary. The ICESCR contains a general limitations clause in Article 4. Limitations of all rights outlined in the ICESCR should be "determined by law and only in so far as this may be compatible with the nature of these rights and solely for the purpose of promoting the general welfare in a democratic society." Accordingly, states have to adopt national laws that incorporate the scope of the limitation. There should furthermore be a legitimate aim, for example, the protection of the public order or security, or the protection of the rights and freedoms of others, and the limitation should be proportionate and appropriate to serve this aim. With respect to the right to science, states may adopt specific measures to limit the conduct of science or the dissemination of scientific results in order to prevent harm to or disrespect for other human rights.[62]

Such specific measures may include free, prior, and informed consent by specific populations, such as indigenous populations or ethnic minorities.[63] In previous chapters, we have seen how the issue of prior consent and the protection in general of these groups have been linked to the democratic access to participate in science policy. This is an important aspect of citizen science. The hope is that citizen participation in science and science policy may highlight the human factor—that is, may help ensure focus on human dignity and integrity, thereby preventing dual use science.[64]

The General Comment also mentions specific groups for whom "temporary special measures might be necessary to achieve substantive equality and remedy current manifestations of previous patterns of exclusion of these groups."[65] Among these groups are women, persons with disabilities, LGBTI persons, indigenous peoples, and persons in poverty. It is significant that LGBTI, indigenous peoples, and persons in poverty have been added to the two other groups, women and people with disabilities, whose perspectives must be integrated into the work of the UN special rapporteur in the field of cultural rights.[66] Even though not directly linked to the issue of dual use, special protection for these groups is necessary since, "in the last decades, the growth of inequalities has undermined the Rule of Law and has had negative effects on the enjoyment of ESCRs," of which the right to science is one.[67]

Whose obligation is it to respect, protect, and fulfill the right of everyone to enjoy the benefits of scientific progress? Whereas the General Comment on science (like other General Comments) mostly concentrates on the obligations that states parties have, the special rapporteur's report and especially the UNESCO Recommendation on Science raise the issue of the responsibility of the individual researcher. I shall come back to the latter in the next subsection. With regard to the right to science, core obligations for states parties that have ratified the ICESCR include respect for the freedoms that are indispensable for scientific research (whose elements are outlined above), promotion of access to the benefits of science and its applications on a nondiscriminatory basis, prevention of harmful effects of science and technology, adoption of mechanisms aimed at aligning government policies with the best available and generally accepted scientific evidence, and strengthening international cooperation, including respect for collaboration of scientists across borders.[68] Key to these core obligations is the notion of progressive realization.

The right to science largely depends on cultural, economic, social, and political development. According to ICESCR Articles 2 and 3, states parties must ensure all rights under the ICESCR equally and without discrimination, just as they must take "deliberate, concrete and targeted" steps toward the full realization of these rights. This principle of progressive realization is one that distinguishes the ICESCR from the other major covenant, the ICCPR.[69]

In General Comment No. 3 of 1990, the CESCR offers an interpretation of the meaning of progressive realization.[70] This concept was included in the ICESCR to acknowledge the fact that states parties are in different economic, social, and political positions.[71] Limited resources are also relevant to the implementation of the right to science since science and technology may be costly. Moreover, in terms of priorities, states may argue that science is not at the top of their list of human needs; even so, they are still legally obliged through different mechanisms to invest, to the maximum possible, in scientific and technological advancement and the sharing of benefits. Furthermore, states parties should start the implementation immediately and should move as expeditiously and effectively as possible toward total realization.[72]

Progressive realization and moving as speedily as possible toward full realization imply that, in principle, the level of protection should not decrease after a certain level has been reached. So-called retrogressive measures are allowed only in exceptional cases. One question with regard to the issue of the realization of the right to science is whether austerity measures and cuts in public budgets for basic science may be categorized as retrogressive measures—and whether states parties in passing such cuts violate their core right-to-science obligation of respecting the freedoms indispensable for scientific research. The General Comment seems to indicate as much, listing as examples of retrogressive measures "the removal of programs or policies necessary for the conservation, the development and the diffusion of science; [and] the imposition of barriers to education and information on science."[73]

Scientific Responsibility and the Precautionary Principle

In Chapter 1, I discussed the link between scientific freedom and scientific responsibility. Along with freedom of science and the enjoyment of the benefits of scientific progress, safeguarding the public from adverse effects of science is an important element of the protection in ICESCR Article 15(1) (b) of every person's right to benefit from science.[74] From a human rights perspective, protecting the public from adverse effects of science thus puts a limit on the scientific freedom enjoyed by scientists.

In addition to the responsibility scientists have toward their fellow citizens, they are accountable to their colleagues within the world of science. When it comes to scientific integrity, most scientists would themselves be the first to admit that certain restrictions on scientific freedom are in order.

Although scientific freedom encompasses the freedom to try out new topics and approaches, the underlying methodology must remain scientifically appropriate. Scientific integrity can be undermined, for example, by forgery and manipulation of data, selection and rejection of unwanted results, manipulation of a representation or an image, and plagiarism and theft of ideas. Falsification of content, unjustified adoption of scientific authorship or coauthorship, and sabotage of research activity through manipulation also are inexcusable acts.[75] If a scientist gets caught in a deliberate act, it will be the end of their career. Catherine Rhodes and John Sulston sum up the responsibility scientists have toward each other and toward society in this way:

> Scientific responsibility includes the responsibilities of scientists towards science and their fellow scientists—doing good science requires, for example, appropriate application of scientific methods, accurate reporting of results, and open dissemination of findings. It is now widely accepted that scientific responsibility extends beyond this and requires some consideration be given to the outcomes and consequences of research. Interpretation and determination of such responsibilities is frequently based on moral considerations.
>
> In relation to the public and decision-makers who influence the direction and application of science and technology—the scientific community has responsibilities because it is in a unique position to present information and knowledge that it is developing about the challenges which face humanity and how they might be addressed.[76]

Scientists have at times themselves called for self-regulation, moreover. In a 2008 article in *Nature*, Paul Berg asks if a meeting similar to the 1975 Asilomar conference on DNA in California could help resolve contemporary controversies concerning the benefits of genetic engineering as well as the risks and ethical dilemmas that it presents.[77] Berg was one of the organizers behind the Asilomar conference who summoned scientists to discuss genetic recombination. Their main motivation was the anticipation of ecological danger, specifically biohazards, posed by advances in genetics. The scientists who attended the conference managed to put into place a system of scientific self-governance, which was largely accepted by the broader scientific community.[78]

There is a lesson in Asilomar for all of science, writes Berg: "The best way to respond to concerns created by emerging knowledge or early-stage technologies is for scientists from publicly-funded institutions to find common cause with the wider public about the best way to regulate—as early as possible. Once scientists from corporations begin to dominate the research enterprise, it will simply be too late."[79] Berg's call to fellow scientists is important, and it is noteworthy that he, like many others approaching science as a public good, draws attention to the potential danger of private moneyed interests taking over. This is a theme that has been touched on already in previous chapters, especially in contexts involving IP.

At the level of science policy, the need for ethical considerations has come up in debates concerning the so-called precautionary principle, which enables rapid response in the face of a possible danger to human, animal, or plant health, or to protect the environment.[80] The precautionary principle has become a fundamental principle in the EU system, allowing decisionmakers to adopt precautionary measures when scientific evidence about an environmental or human health hazard is uncertain and the stakes are high.[81] There is no universally accepted definition, but the one proposed by UNESCO is repeated in the General Comment on Science.[82]

Application of the precautionary principle has proven very controversial. Many scientists complain that if it is applied too broadly, it may make scientific progress impossible. Others counter that we can no longer rely on scientists to self-regulate and to uphold the principle of science as a public good. Private funding and commercial interests have become so ubiquitous that we need the precautionary principle. However, just as there is no agreement on definitions, experts and institutions argue about methods for hazard assessment and about when to apply precautionary measures (cost-benefit analysis, risk trade-off analysis, cost-effectiveness analysis, pros and cons analysis of action and inaction, etc.).[83] Reconciling the defense of scientific freedom with the interest of the public in protection from dual use science and technology is truly a balancing act.

Conclusion

"Human beings have always tried to systematize and explain what exists and happens around them," wrote Brøndegaard in the early 1970s. "The 'mysterious' forms, colors, properties, behavior etc. of animals and plants have made human beings, who are born curious, ask the question 'why'? This is how etiological legends involving nature have become by far the most numerous kind."[84] Such explanation legends, as Brøndegaard called them, could focus on a name, for example, whose origin and meaning were unknown and about which people speculated and let their fantasies run wild.

The explanation legends concerning the elderberry plant are an interesting example. *Flora Danica* plate 545 of 1771 displays *Sambucus nigra*, or elder. When in bloom, elder flowers release a strong, sweet smell, and both the flowers and the berries have been utilized "for making juice, wine and as a flavour additive in desserts and in cooking generally."[85] Children have used twigs from elderberry shrubs to make toys, and elderberries and flowers also figure in a variety of herbal medicines—as well as in numerous folk legends. Hans Christian Andersen's 1844 fairy tale "The Elder-Tree Mother" builds on these legends.[86]

Sambucus ebulus or dwarf elder is another elder species found in Denmark. Its name in Old English is walewort, but it has also become known in some places in England as danewort, a name traceable to the belief that it sprang from the blood of Danes killed in battle. Another account of its English name has it that the plant was brought to England by the Danes and was planted on the battlefields and graves of their slain compatriots.[87] English theologian and naturalist John Ray (1627–1705) is among those who mentioned this account. His work was familiar to Linnaeus, who, transferring the *Sambucus ebulus* explanation legend to Sweden, wrote in 1741 that dwarf elder may be found around the city of Kalmar, where "common people believe that it grows from the blood of dead [Danes] killed in war."[88] The story about flowers growing from blood is familiar from other parts of the world, perhaps most famously in connection with the poppies growing near one of Flanders Fields' mass cemeteries in Belgium during and after World War I.

With regard to Linnaeus's role in spreading this particular explanation legend, Brøndegaard comments, it is remarkable that "an otherwise so rational naturalist as Linnaeus consciously contributed toward the dispersal of the explanation legend [of *Sambucus ebulus*] by bringing it to Sweden without revealing where he took it from."[89] Combining natural with cultural knowledge, Linnaeus acted much like an ethnobotanist. He also plagiarized someone else's work. Not much damage was done in this case, we may speculate; the offense on Linnaeus's part does not amount to anything close to dual use science. Yet, one of the reasons why plagiarism is considered to be a very serious academic crime is that it jeopardizes the trust in scientific integrity—and in the importance of scientific freedom. Over the past few decades, the public has become more aware of scientific misconduct, and interest in the importance of ethical standards in the conduct of science has grown.

As we have seen in this chapter, reconciling the search for scientific truth with the needs for protection against dangerous applications resulting from this search and scientific misconduct is a balancing act.[90] The scientific freedom promised in ICESCR Article 15(3) may on occasion be at odds with the rights of the public outlined in the other parts of Article 15. Since the 1980s, academic and funding agencies have accordingly issued public statements and guidelines, as well as procedures for dealing with allegations of misconduct in science—a topic further explored in Chapter 5.[91] In the EU system, the concept of responsible research and innovation has been designated a "crosscutting issue" and is now implemented in all research framework programs to ensure that scientists and their collaborators take into consideration the social, ethical, cultural, and environmental consequences of their research.[92]

The General Comment reflects this. On the one hand, it follows the suggestion of the special rapporteur that the RtS should be read together with the right of the people to take part in democratic decisions—and that in the end, the public should decide: "In controversial cases, participation and transparency become crucial because the risks and potentials of some technical advances or some scientific researches should be made public in order that society, through an informed, transparent and participatory process, can decide whether or not the risks are acceptable."[93] In the subsection on limitations, however, the comment emphasizes how "nevertheless, any limitation on the content of scientific research implies a strict burden of justification by States, in order to avoid infringing freedom of research."[94] Here, as throughout the comment, we see how the wording of the four parts of Article 15 provides the public and other decisionmakers with a helpful tool for recognizing various stakeholders and for making sure their interests are understood and heard before decisions are made that affect us all.

CHAPTER 5

The Right to Science and International Cooperation and Solidarity

> The States Parties to the present Covenant recognize
> the benefits to be derived from the encouragement and
> development of international contacts and co-operation
> in the scientific and cultural fields.
>
> —Article 15(4) International Covenant on Economic,
> Social and Cultural Rights

Science and scientific scholarship cannot exist without international cooperation and knowledge exchange. Article 15(4) is therefore, I argue in this chapter, a starting point or prerequisite for any discussion concerning the right to science.

In March 2020, the WHO declared coronavirus disease (COVID-19) a pandemic.[1] The United Nations Development Programme talked about the worst global health crisis since World War II[2]—and the CESCR quickly followed suit by stating that "the COVID-19 pandemic is a global crisis, which highlights the crucial importance of international assistance and cooperation."[3] Responses to the pandemic ought to reflect the best available scientific evidence, the CESCR argued with reference to its newly adopted General Comment No. 25 on Science, pandemics being "a crucial example of the need for scientific international cooperation to face transnational threats."[4]

Furthering international cooperation is a core principle of the ICESCR, which figures prominently in three of the covenant's articles. Article 2(1) mentions international assistance and cooperation in the context of the progressive realization of economic, social, and cultural rights, and Article 11, concerning the right to an adequate standard of living, likewise highlights

the importance of international cooperation. Finally, Article 15(4) emphasizes "the benefits to be derived from the encouragement and development of
international contacts and co-operation in the scientific and cultural fields."[5]

During the coronavirus crisis, scientists around the world have worked
toward finding appropriate vaccines for and treatment of the coronavirus.
Most of them are trained in modern, Western medicine, but traditional
medicine has also come into play. Some scientists have suggested the use of
traditional Indian medicinal herbs as a possible therapeutic approach, for
example.[6] Others have pointed to traditional Chinese medicine as a source
of hope not only for the treatment of COVID-19 but also for the prevention
and control of the coronavirus.[7] If clinical trials are involved and if these traditional medicinal approaches are "based on critical inquiry and open to falsifiability and testability," they can be scientifically validated.[8] In this way, the
COVID-19 pandemic offers a good example of the possibility of establishing
links between the empirical knowledge of traditional medicines and modern
Western science, suggested by ethnobotanists and others (see Chapter 4).

The relationship, and often tension, between local, traditional knowledge
and modern, scientific knowledge production is especially pertinent to the
exploration of what is and is not considered to be science, as we saw in Chapter 4. This relationship has been relevant to themes covered in other chapters—
and will come up again in this chapter, whose focus is the historical, legal, and
conceptual meanings of Article 15(4). This last part of Article 15 relies on, and
only makes sense when read together with, the article's first three parts. In the
current context of pandemics and climate change, global in scope like the interdependence of the world's economies, it is by no means the least significant.

The making of the *Flora Danica* herbarium involved both local and international knowledge exchange. In the first part of the chapter, I review it from
the perspective of botany as "colonial" knowledge exchange in the eighteenth
century.[9] Scholars interested in this period have explored the relationship
between botany and European colonial expansion with its problematic implications for local peoples and environments.[10] To the traditional celebration
of science's inevitable progress during the Enlightenment as an international
force, they argue, should be added illustrations of how the fight to establish
scientific knowledge as valuable was linked with commercial and imperial
exploitation, suppressing local identities and traditions.[11] As the *Flora Danica*
project never included plants from the tropical Danish colonies, the issue
of "colonial" knowledge production is primarily relevant with regard to the
northern dependencies of the Danish realm, especially Greenland.

The *Flora Danica* project aimed at identifying the native plants and their usefulness, as Georg Christian Oeder wrote to prospective subscribers in 1761. But a number of the plants growing wild in the various parts of the realm had actually come from other parts of the world.[12] Some of them have since been reduced in number or have been displaced by invasive plants. Today, when humanity's impact on the earth is so profound that scholars talk about a new geological epoch—the Anthropocene—it is even more unclear what the "natural" or "native" flora is. Global traveling and trade, changes in land use, cultivation, and spraying with chemicals have caused planetary changes on an unprecedented scale. Directly as well as indirectly, the story of the *Flora Danica* project allows for reflection on the meaning of native versus traveling, or invasive, plants and on the dynamic quality of the natural world in general.

The second part of the chapter explores the legal and conceptual meaning of ICESCR Article 15(4). International human rights law has developed through states entering into treaties, governed by the international law of treaties as articulated through customary international law and the Vienna Convention of the Law of Treaties from 1969.[13] Some treaties emphasize the rights of everyone living within a particular country, including foreigners, whereas a few list protections that go beyond national borders. Among the latter are the human rights outlined in the ICESCR, mentioned above. A survey of relevant recent human rights instruments such as the 2020 General Comment on Science and the 2017 UNESCO Recommendation on Science and Scientific Researchers will allow us to establish—or at least approach an understanding of—what international cooperation means with respect to the right to science.

The third and last part of the chapter continues the discussion from Chapter 4 concerning bioprospecting and its consequences for local groups and environments. If done with respect and concern for the traditional knowledge of plants that is often involved, bioprospecting—the systematic search for useful products derived from bioresources including plants and the further development of these plants for commercialization—can benefit all stakeholders and society in general.[14] Intellectual property (IP) plays a prominent role when it comes to the global transfer of knowledge, as we have also seen in previous chapters. Here, IP is explored from the perspective of international cooperation and assistance, voiced as an imperative in the ICESCR and other major human rights instruments. The issue of scientific (mis)conduct and the importance of ethical standards in the conduct of science is also relevant in this context and will be explored toward the end of the chapter.

Flora Danica and Colonial Botany

"With its dialectical ability both to upset and to objectify social relations," argues Staffan Müller-Wille, science "constitutes one of the motors of colonialism writ large, rather than merely serving as one of its instruments."[15] To Müller-Wille and other historians of science, Linnaean taxonomy provides a good example. Though Sweden had no major colonies, Linnaeus's binomial nomenclature proved to be instrumental for colonial conquest. Designed as it was to ignore local features and to mediate between, or work at a level above, different cultures, binomial nomenclature made global botanical exchange possible by "reducing plants to specimens, numbers, and names so that a specimen, once identified, represented any plant of its type anywhere in the world where it is found."[16]

For Oeder, following Buffon and von Haller, as we saw in Chapter 4, the problem with Linnaean taxonomy was its artificiality and failure to take into account the natural variation within and hybridization between species. In a certain sense, the criticism voiced by Müller-Wille, Londa Schiebinger, and others also concerns the artificiality of the Linnaean system, the moving away from particularized knowledge to universal systematizing. As they see it, in aiming to be globally useful this system omits or suppresses something they consider very relevant, namely local knowledge, culture, and social relations.

Colonial botany—"the study, naming, cultivation, and marketing of plants in colonial contexts"—was arguably "born of and supported European voyages, conquests, global trade, and scientific exploration."[17] In his 1761 presentation of and invitation to subscribe to a new work to be called *Flora Danica*, Oeder mentioned that this work would "account for all the plants that grow wild and without the help of humans in the two kingdoms of His Majesty and in his other lands in Europe."[18] The two kingdoms were Denmark and Norway, and the "other lands in Europe" were the Faroe Islands, Iceland, and Greenland, as well as parts of what is today northern Germany.

As we have seen in previous chapters, the idea was to describe plants that could be of immediate use to Danes.[19] "It is the will of His Majesty," wrote Oeder in the invitation to subscribe, "that with the work it should both be possible to identify the native plants, and that their usefulness to the inhabitants of these kingdoms and countries should be explained."[20] Martin Vahl (1749–1804), editor of *Flora Danica* between 1787 and 1799, did publish a book on plants he had received from what was then known as the Danish West Indies, but these plants did not become a part of the *Flora Danica* itself.[21]

Neither did the flora described by Moravian missionary Christian Olden-dorp (1721–1787), who visited the Danish settlements for eighteen months between 1767 and 1769. Oldendorp, who had a background in science, was sent from Germany to the West Indies to write a report on the Moravian mission in the islands. His report was published in 1777.[22]

The historic link between botany, religion, and missionary work has been of special interest to scholars of natural history and global expansion.[23] In the Danish context, the Protestant Moravian Church, which was founded in the eighteenth century but traces its origin to fifteenth-century movements in Bohemia and Moravia, played a small but remarkable role. On December 10, 1771, King Christian VII (1749–1808) issued permission for the Moravian Brethren to found a congregational city in Denmark. This happened at the behest of Carl August Struensee (1735–1804), who had been called to Denmark in 1771 to run the financial services of the kingdom by his younger brother, Johann Friedrich Struensee (1737–1772), the king's German-born court physician who served as the kingdom's de facto ruler between 1770 and 1772. As we saw in Chapter 1, the younger Struensee was beheaded for allegedly having plotted to kill the king. Not so his brother, who was allowed to go back to Prussia, where he had a distinguished career as minister of finance. The new Moravian congregational city was founded in mainland Jutland and named Christiansfeld in honor of King Christian. In 2015, Christiansfeld was added to the UNESCO World Heritage List.[24]

What had impressed both Struensee brothers was the importance the Moravian Brethren gave to artisanal skill, both in the mother community in Herrnhut, Germany, and in the missions founded around the world with the aim of converting natives. Writes Michael T. Bravo, "The Moravian Brethren emphasized practical, artisanal skills, though the missionaries were in fact self-selecting volunteers. Curing illness and finding reliable sources of food were integral to the planning and realization of conversions. Hence the skills required to name flowers, vegetables, trees, and animals and to cultivate them, harvest them, and transform them into useful forms for everyday life were highly valued."[25] The first foreign Moravian mission of what was to become a worldwide Christian outreach program was to the West Indies in 1732. Within a couple of decades, missions were also established in Greenland, Suriname, South Africa, and Algiers and among the North American Indians.[26]

Moravian mission work was begun in Greenland in January 1733.[27] The first years were not easy for the first three missionaries. Finding enough food was difficult enough, but they also had to learn the local language to get by.

Still, the group of missionaries grew. In 1736, four new missionaries arrived from Herrnhut, three of whom were women. Early on, one of the Moravians got involved in a theological dispute with Danish Lutheran missionary Paul Egede, the son of Hans Egede, who had arrived in 1721 to missionize among the Inuits. Paul Egede stayed in Greenland for four years, and he was the first to collect Greenlandic plants for a small herbarium (1739).[28] His dispute with the Moravians stopped only when he left Greenland in 1736. As for the Moravians, toward the end of the nineteenth century they had obtained a solid presence, but poor financial conditions forced the Moravian Church to cut down on its missionary activities. The last Moravian missionaries left Greenland in September 1900.[29]

Functioning gardens were essential throughout the 167 years that the Moravian missionary experiment lasted in Greenland. Mastering the use of local flora therefore "involved the skills of carpenters, tailors, cobblers, blacksmiths, cooks, and gardeners—as well as indigenous cultivators and harvesters," writes Bravo.[30] This is reflected in David Cranz's 1767 *History of Greenland*, which included a small selection of plants.[31] A Protestant missionary monograph and a hybrid of religious and travel book, Cranz's work established a literary model that was followed by, among others, fellow Moravian Christian Oldendorp when he published his 1777 Caribbean report, mentioned above.[32]

When describing the local flora, Cranz referred to the various ways in which it was used by the Inuits. One example is the practice of insulating stone houses with moss. Another concerns the kayak. Cranz was impressed both with the skills that went into its design and with the expertise required to use a kayak in the hunting and fishing economy for sustenance. Though the indigenous population was portrayed with respect, Cranz, Oldendorp, and other Moravian naturalists tended to "domesticate" local competences such as kayaking— perceiving them as being of interest only as they related to their own project.[33] The result was, Michael Bravo argues, that the indigenous population was "too often stripped of the ownership of their labor. . . . Artisanship was in fact an instrument of missionary ideology, as well as a category of self-understanding, which helped to justify the Moravians' ownership of slave plantations in the West Indies and the transformation of hunting cultures such as the Inuit of Greenland."[34]

With regard to collecting plants in what Oeder called the king's "other lands in Europe," it was Jens Vahl (1796–1854), son of Martin Vahl, who, during an exploratory journey to the east coast of Greenland between 1828 and 1831, succeeded in assembling the first systematic collection of plants in Greenland.[35] This collection was put to good use by later botanists, notably Johan M. C. Lange (1818–1898), who took over as editor of *Flora Danica* in 1858 and successfully completed the work in 1883. That same year he also published four collections of *Flora Danica* plates, each of which had an accompanying text. One of these was *Icones Florae Groenlandicae* with 336 plates of illustrations of Greenlandic plants. The other three consisted of plates from *Flora Danica* that portrayed various Scandinavian plants, both medicinal and agricultural.

Plants were furthermore included from Iceland. The Danish Royal Society for Arts and Science had sponsored a journey of exploration in Iceland, undertaken by two young, well-educated Icelanders between 1752 and 1757. The account that was published afterward offered descriptions of the flora of Iceland. Oeder also had an expedition sent to Iceland in 1765–1766 to collect plants specifically for *Flora Danica*. And as for the Faroe Islands, they "were not seen as sufficiently different from Norway and Iceland to warrant special expeditions."[36] About 650 plants were included in *Flora Danica*, though most of them were also known from other parts of the Nordic countries.

Native, Invasive, or Hybrid?

Though scientific research is typically embedded in a national setting, it is inherently international. Creativity and inspiration know no borders. Neither does building on the understanding of others to make intellectual progress. Scientists concentrating on their own local settings without being attentive to what is happening at the international level risk producing scholarship that is provincial and uninteresting. This is what happened in Denmark during the Enlightenment period. "To our eyes," wrote English clergyman and mineralogist Edward Daniel Clarke, who visited Copenhagen in 1799 on his way to Norway and Sweden, "it seemed indeed, that a journey from London to Copenhagen might exhibit the retrocession of a century; every thing being found, in the latter city, as it existed in the former a hundred years before. . . . In literature, neither zeal nor industry is wanted: but compared with the rest of Europe, the Danes are always behind in the progress of science."[37]

Whatever valuable natural science research did take place at the time in Denmark, argues science historian Helge Kragh, was indebted to foreign experts and contact with scientific centers of excellence in other countries, especially neighboring Sweden and Germany.[38] Kragh highlights as a prime example of such a foreign expert (only grudgingly allowed into the country) Georg Christian Oeder, who, by initiating the *Flora Danica* project, made a considerable contribution not only to Danish natural history but also to the public health of the country.[39] The illustrators of the first few fascicles, it bears repeating, were also foreigners. Oeder persuaded engraver Michael Rössler (1705–1777) and his son Martin (1727–1782) to leave their native Germany for Copenhagen in order to work with him on the *Flora Danica* project.

The plants found, described, and illustrated by Oeder, Michael and Martin Rössler, and their successors were the result of thorough field explorations and provide the earliest calculation we have of the number of plants growing within the Danish realm. *Flora Danica* contains illustrations of about 1,400 species and varieties of plants to be found within what is today the Kingdom of Denmark. The work did not contain any assessment of whether each species was native or had been introduced into the realm.[40] It was two of Oeder's successors, Jens Wilken Hornemann (1770–1841) and Johan M. C. Lange, editors of *Flora Danica* from 1805 to 1841 and 1858 to 1883, respectively, who first attempted to evaluate whether plants were native—and if not, how they could have come to grow in Denmark.[41] Both published handbooks or systematic descriptions of Danish plants, which contained surveys of plants introduced to or not naturally growing wild in Denmark. Whereas it was comparatively easy, Lange thought, to establish which kinds of plants had been introduced on purpose by human beings (e.g., to grow in fields or gardens), it was difficult to differentiate between originally wild-growing plants and plants introduced in ancient times or during the Middle Ages.[42]

In a publication from 1839–1840, Hornemann counted about three hundred introduced or naturalized plants; a half century later, Lange raised that number to about four hundred.[43] Today, botanists and others increasingly refer to invasive plants or even invasive alien species, the latter defined by the European Commission as "animals and plants that are introduced accidentally or deliberately into a natural environment where they are not normally found, with serious negative consequences for their new environment."[44] While not all alien species are a cause for concern, "a significant subset of alien species can become invasive and have serious adverse impact on biodiversity and related ecosystem services, as well as have other social and economic impact, which should be prevented. Some 12,000 species in the environment of the Union and in other European countries are alien, of which roughly 10 to 15 percent are estimated to be invasive."[45] Debates involving invasive plants can very easily become politically sensitive—especially when these plants are implicitly compared to other, essentially unrelated globalization phenomena such as migration.[46] Some scholars have therefore suggested the use of other terms—terms that come with less ideological baggage. The idea behind the research project and scientific flora Flora Nordica, for example, is "to work out a scientific taxonomic and biogeographical treatment of wild, naturalized, and casual plants" found in the Nordic countries.[47] Written in English, Flora Nordica was started in 1987 and both professional and amateur botanists have taken part.[48]

Many countries have established lists of invasive as well as threatened wild-growing plants. These lists are updated regularly and often follow the guidelines established by the International Union for Conservation of Nature (IUCN).[49] Sometimes, it is interesting to look at what is *not* on these lists. Going back to the Danish context, why is it, for example, asks Friis, that ramsons (*Allium ursinum*) is not placed on the list of invasive plants?[50] Illustrated in *Flora Danica* plate 757 (1778) and today an often-used ingredient in New Nordic cuisine, ramsons is a wild relative of chives that is sometimes also called wild garlic. In *Flora Danica* it was known only in certain parts of Denmark, but today it not only is common all over the country but also seems to be aggressively replacing other plants.[51]

Even if there may thus be a certain element of arbitrariness involved in the lists of invasive or invasive alien species, most botanists would not question the need for such lists. In the age of the Anthropocene, humans now dominate the planet to such an extent that indigenous wildlife is seriously endangered. The lists of invasive and threatened plants amply document this.

To British biologist Chris D. Thomas, however, current assessments are altogether too pessimistic. As the title of his 2017 book, *Inheritors of the Earth: How Nature Is Thriving in an Age of Extinction*, indicates, Thomas is interested in Anthropocene evolution—and in why some species decline and disappear while others are successful.[52] In his book, written for a general audience, he argues that some species *are* successful, that biodiversity gains *have* happened, and that formation of new species through either hybridization or speedy evolution and adaptation is a defining characteristic of the Anthropocene. "A global-scale spate of rapid evolution," he writes, "is in full flow. . . . The Earth is poised for a massive acceleration in the formation of new species."[53] Thomas's arguments are refreshingly optimistic. Some critics have pointed out that they are also somewhat oversimplifying, the result being a misrepresentation of what is considered to be basic knowledge in nature conservation and ecology.[54]

Thomas's book and the criticism it has received provide a good example of the difficulty of writing popular science books—of disseminating science, as outlined in Article 15(2) and discussed here in Chapter 3. In pointing to Anthropocene evolution as an inevitably global phenomenon, moreover, Thomas also points to the way in which all parts of the world are today interwoven. Whether people want this to be the case or not, what happens in one locality will have ramifications for what happens at the other end of the globe. This global interdependence is reflected in international human rights law.

ICESCR Article 15(4), Legally and Conceptually

International human rights law is governed by the international law of treaties. Some treaties emphasize the rights of everyone living within a particular country, citizens as well as noncitizens. According to the 1966 International Covenant on Civil and Political Rights, for example, "each State Party to the present Covenant undertakes to respect and *ensure to all individuals within its territory and subject to its jurisdiction* the rights recognized in the present Covenant" (emphasis added).[55] The 1950 European Convention on Human Rights also secures "*to everyone living within [the] jurisdiction [of the High Contracting Parties]* the rights and freedoms contained in Section 1 of this Convention" (emphasis added).[56] Human rights protection, as Mark Gibney reminds us, is not based on citizenship, and a state's human rights obligations do not end at its own territorial boundaries.[57]

A couple of international human rights instruments explicitly include provisions for human rights that are international in scope. One example is the 1989 Rights of the Child Convention.[58] Another is the ICESCR, whose Articles 2, 11, and 15 all mention the importance of international cooperation and assistance. According to CESCR General Comment No. 3 (1991), because it describes the nature of the legal obligations that states parties to the covenant have, Article 2 is of "particular importance to a full understanding of the ICESCR and must be seen as having a dynamic relationship with all of the other provisions of the Covenant."[59] One of the most important of these obligations concerns what has come to be called the obligation of "progressive realization," outlined in Article 2(1)—namely that states must move as expeditiously and effectively as possible toward the full realization of all economic, social, and cultural rights (see Chapter 4). Importantly in our context, this is an obligation that refers not only to domestic assets but also to resources available through international cooperation and assistance.[60]

States have a duty to cooperate with other states on development issues, but also with regard to other economic, social, and cultural rights. Moreover, helping "is particularly incumbent upon those states which are in a position to assist others in this regard."[61] A certain extraterritorial or international scope seems to have been intended from the beginning. When the ICESCR was being drafted, for example, René Cassin suggested that, "by providing for recourse to international cooperation instead of allowing the enjoyment of rights to be put off, [the reference to international cooperation] filled the gap between what States could in fact do and the steps they would have to take to meet their obligations under the Covenant."[62]

In concrete terms, it is still somewhat unclear what precisely the extraterritorial reach is of the ICESCR and what it entails for states parties and others.[63] Since the publication of General Comment No. 3, CESCR has attempted to develop the scope and content of ICESCR Article 2(1) in various general comments and other documents.[64] With regard to the right to science, General Comment No. 25 offers especially relevant guidance.[65] The comment devotes one entire section (out of seven) to international cooperation. Consisting of eight paragraphs, this section opens with a strong call for international solidarity: "States need to take steps through legislation and policies, including diplomatic and foreign relations, to promote a global environment favorable for the advancement of science and the enjoyment of benefits from its applications."[66] The following four paragraphs outline the various international dimensions of states' obligations.

The first dimension concerns the obligation of states to make it possible for scientific researchers to travel and to share their research with colleagues around the world—a topic to which I will come back in the next section. The second and third dimensions relate to the disparities between countries when it comes to scientific research and its technological applications. Developed countries have obligations to "contribute to the development of science and technology in developing countries" in a variety of ways, for example, by allocating resources to improving scientific education, research, and training. But developing countries also have obligations. They should "resort to international assistance and cooperation" if complying with their obligations vis-à-vis the ICESCR is outside their financial and technological reach.[67]

Referring here as elsewhere to the 2017 UNESCO Recommendation on Science and Scientific Researchers, the comment also seems to be inspired by arguments such as the ones made by founding father Cassin, quoted above. An important aspect of the second dimension is the problem of brain drain. The comment warns developed states against furthering brain drain, as this can have dire effects for developing countries.[68] The third dimension concerns the importance of sharing the applications and benefits resulting from scientific progress across the world. The developed world must allow poor communities and vulnerable groups to benefit from its scientific research and its concrete output, "especially when these benefits are closely related to the enjoyment of ESCRs."[69]

The fourth and final dimension of states' obligations relates to the fact that the most severe problems today, such as climate change and the loss of biodiversity, are global problems that need global solutions. Likewise, dual use science and technology involving, for example, artificial intelligence and the threat of weapons of mass destruction can be countered only by means of international cooperation.[70] A further example is pandemics. Without directly referring to COVID-19, the comment underscores the importance of "sharing the best scientific knowledge and its application, especially in the medical field" during as well as after a pandemic.[71]

Finally, states have extraterritorial obligations with regard both to monitoring the conduct of multinational companies and to realizing the right to science fully. States should ensure that multinational companies respect the right to science, and if people fall victim to the activities of such companies, states must come up with remedies, including legal ones.[72] This monitoring obligation will no doubt become increasingly important as global technology companies amass ever more power in the digital economy. In terms of the full realization of the right to science, IP rights are key to ensuring that traditional knowledge is properly credited and protected. States should therefore make sure that IP regimes "foster the enjoyment" of the right to science in such a way that not only scientific knowledge but also the benefits arising from its application are equally shared.[73]

International Mobility and Freedom of Movement

The international cooperation and mobility emphasized in the General Comment on Science is challenged in various ways. Scholars at Risk (SAR) is an international network of institutions and individuals working for the protection of scholars and the promotion of academic freedom. Based at New York University, one of its core activities is the Academic Freedom Monitoring Project, which "investigates and reports attacks on higher education with the aim of raising awareness, generating advocacy, and increasing protection for scholars, students, and academic communities."[74] According to this project, all is far from well when it comes to the international mobility, academic freedom, and freedom of movement of scientists. Between August 2020 and August 2021, the project has tracked incidents worldwide involving seventy-seven killings, acts of violence toward, or instances of disappearance of scholars; seventy-one wrongful imprisonments or detentions; twenty-nine wrongful prosecutions; twenty-four losses of position; four travel restrictions; and thirty-three other significant events.[75]

The SAR numbers form a bleak background commentary to the significance placed in the General Comment on scientific freedom of movement—the ability of scientists to move freely between places of employment and residence, without undue influence from governmental actors. Freedom of movement for scientists also features prominently in the 2017 UNESCO Recommendation on Science and Scientific Researchers, which urges member states to actively "promote the interplay of ideas and information among scientific researchers throughout the world" and to "facilitate [the travel of these researchers] in and out of their territory."[76] This right is important because "the interplay of ideas and information among scientific researchers throughout the world . . . is vital to the healthy development of the sciences."[77]

Underlying much of the recommendation, however, is a wish (or perceived need) to balance the freedom of movement of scientific researchers against the fear of brain drain, as briefly mentioned above. Article 11 advises member states to attempt to create a "general climate" that is supportive of scientific researchers, for example.[78] The emphasis is on "sufficient attraction" to and "confidence" in the "career" or "vocation" of science, reflecting a worry that the scientific career is not glamorous or rewarding enough to do justice to its importance and therefore may fail to attract sufficient interest among young people to "maintain a constantly adequate regeneration of the nation's pool of scientific researchers."[79]

The worry about brain drain is made explicit in the preamble: "*Conscious* that the phenomenon frequently known as the 'brain drain' of scientific researchers has in the past caused widespread anxiety, and that to certain Member States it continues to be a matter of considerable preoccupation; having present in mind, in this respect, the paramount needs of the developing countries; and desiring accordingly to give scientific researchers stronger reasons for serving in countries and areas which stand most in need of their services."[80]

Brain drain is the tendency of the most highly educated or most skilled of a country's population to leave for employment elsewhere, perhaps after finishing an education at home, or alternatively the tendency for foreign students to stay in the countries in which they studied because there are no long-term prospects back home. Brain drain is highly controversial, because the countries experiencing it typically are those most in need of homegrown scientific talent. A phenomenon similar to brain drain, though it is not referred to as such, is the failure of public institutions to hold on to skilled scientists. Salaries and the possibility of earning considerable sums of money from patents and shares make private companies attractive to these scientists. This means that public services may suffer and that the public ends up having to pay for benefits that might otherwise have been freely available.

Brain drain gives rise to a natural tendency to want to implement policies to retain scientific talent. Such policies come in widely different forms. Some are intended to make the pursuit of science and a scientific career in a scientist's home nation so attractive as to prevent that scientist from leaving. This seems to be the intention behind Article 11 of the recommendation, as we just saw. Likewise, Article 18 mentions various ways in which member states may improve conditions for scientific researchers by helping them benefit from the international dimension of research and development. Establishing partnerships between scientific communities in order to further bilateral and multilateral agreements may allow developing countries to expand their capacity for scientific research, for example. Likewise, ensuring equal and fair access to scientific knowledge may make the sharing of data and educational resources easier across national borders.[81]

As member states are doing their very best to encourage scientists to stay put, they must take care not to violate the scientific freedom of these scientists. While it is acceptable for states to make the domestic scientific scene attractive enough to retain talent, it would be unacceptable for them to prevent the free movement of scholars into and out of their territories. This would have intolerable consequences for the pursuit of knowledge and would not be compatible with the nature of the rights outlined in the ICESCR. The line between efforts to hold on to talent and violating scientists' freedom of movement is therefore a very fine one.

Elsewhere in the recommendation, the related issues of ethical review and scientific integrity come up. This is the case in Article 20, which urges states to make sure that transnational research and development activities are conducted with respect for human rights, especially when human subjects are involved.[82] In order to draw the attention of scientists themselves to possible ethical issues arising from their research, moreover, their basic training has to cover intellectual integrity and the human, civic, and ethical implications as well as the probable and possible social and ecological consequences of scientific research.[83] The implication throughout is that human rights may work as a kind of increasingly necessary ethical check on scientific research.

The 2017 recommendation updates and supersedes the 1974 UNESCO Recommendation on the Status of Scientific Researchers. Interestingly, the more recent recommendation adds "the scientific method" and "sensitivity to conflict of interest" to the list of techniques for awakening and stimulating desired personal qualities and habits of mind in the 1974 recommendation. One reason for revising the latter was the perceived need to "provide a powerful and relevant statement of science ethics as the basis for science policies that would favour the creation of an institutional order conducive to the realization of Article 27, paragraph 1 of the Universal Declaration of Human Rights."[84] The direct mention here of UDHR Article 27(1) is significant. The 2017 recommendation was one of the first UN documents to display everyone's right "to share in scientific advancement and its benefits" so prominently and to call it a common good.[85]

In sum, the international obligations of states relate to the sharing across countries of scientific knowledge and its applications in such a way that all communities benefit. The publication of both the 2017 UNESCO recommendation and the 2020 General Comment on Science underscores the fact that science and technology are key to ensuring that "no one is left behind" in the shift to a sustainable path, as pledged by the 2030 Agenda for Sustainable Development, adopted at the UN in 2015.[86] Science ethics and the urgency of avoiding dual use science and technology are highlighted, as is the relevance of a human rights approach. In terms of the individual researcher, the many incidents tracked by the SAR Academic Freedom Monitoring Project demonstrate the need for transnational action to protect scientists, as science (the truth?) regrettably is perceived in various parts of the world as a political (and manipulable) liability.

Bioprospecting, Transfer of Knowledge,
and Intellectual Property

The last two parts of Article 18 of the 2017 Recommendation touch on the issue of IP. Member states should ensure protection and credit for contributions to scientific knowledge, including those involving traditional knowledge. Member states should furthermore do their utmost to prevent biopiracy and illegal trafficking of organs and other tissue just as they should work toward defending human dignity and the privacy of personal data.[87]

The General Comment on science also reflects the way in which IP and other ownership-related issues are central to discussions involving bioprospecting and the global transfer of knowledge in general. In order to further what the comment calls our "global intercultural dialogue for scientific progress," we cannot afford to lose input from anyone. This includes the voices of women and the disabled, to take two groups whose perspectives the mandate for the UN special rapporteur in the field of cultural rights asks the mandate holder to integrate into their work (see Chapter 2). Such an intercultural dialogue is also pertinent to local communities around the world whose knowledge about nature may be especially relevant in today's battle against climate change. With regard to these groups, it is furthermore essential that "science should not be used as an instrument of cultural imposition." Along with other measures, special IP regimes that ensure the ownership and control of traditional knowledge should therefore be developed.[88]

Environmentalists and NGOs first introduced the term "biopiracy" in the early 1990s to refer to bioprospecting gone wrong. Like bioprospecting, biopiracy involves the identification and commercialization of biological resources such as plants. But whereas the former, some argue, can be done with respect for local knowledge and for the benefit of all, biopiracy is associated with the illegitimate use and appropriation, through patents and other kinds of IP, of local knowledge by foreign commercial actors, typically Western biotechnology companies.[89] What is being appropriated is not just the product—say, a particular plant—itself, but also the knowledge that goes into the uses of that product. In IP language, biopiracy involves both tangible and intangible capital and assets, or just tangibles and intangibles.[90]

Biopiracy has come to be seen by some critics as a kind of extension of colonial practices such as the ones I discussed above in connection with colonial biology.[91] The rising levels of IP protection have added an extra layer to these practices. As a result especially of the Trade-Related Aspects of

Intellectual Property Rights agreement, which came into force in 1995 and imposed on all contracting members of the WTO minimum standards of protection for several areas of IP, as we saw in Chapter 2, biopiracy has been turned into an issue of global justice.[92] From the perspective of local communities, the key issue is that Western scientists and companies benefit from their knowledge and practices, whereas the communities themselves do not. For the past few years, a number of these local communities have therefore started to fight for the ownership of their traditional knowledge and practices by means of IP. The problem is, though, that concepts of individual ownership and innovation, which are central to patents and other kinds of IP, are alien to many local communities and their notions of communal ownership passed on orally from generation to generation.[93]

The launch in 2007 of the United Nations Declaration on the Rights of Indigenous Peoples (UNDRIP) marked an attempt to draw international attention to the importance of IP issues for traditional communities.[94] Article 31 has the following wording:

> 1. Indigenous peoples have the right to maintain, control, protect and develop their cultural heritage, traditional knowledge and traditional cultural expressions, as well as the manifestations of their sciences, technologies and cultures, including human and genetic resources, seeds, medicines, knowledge of the properties of fauna and flora, oral traditions, literatures, designs, sports and traditional games and visual and performing arts. They also have the right to maintain, control, protect and develop their intellectual property over such cultural heritage, traditional knowledge, and traditional cultural expressions.
>
> 2. In conjunction with indigenous peoples, States shall take effective measures to recognize and protect the exercise of these rights.[95]

Article 31 is markedly different from the ICESCR in that it openly mentions IP rights. The wording of ICESCR Article 15(1)(c) is that everyone has the right "to benefit from the protection of the moral and material interests resulting from any scientific, literary or artistic production of which he is the author." In her three reports on the right to science, copyright, and patents, UN special rapporteur in the field of cultural rights Farida Shaheed specifically insisted that IP rights are not human rights and that authors' rights are different from IP rights (see Chapter 2). However, the 2017 UNESCO Recommendation on Science and Scientific Researchers lists the right of individual researchers to be remunerated for their work as an IP right.[96]

In the context of both UNDRIP and the 2017 UNESCO recommendation, I speculate that underlying historical reasons and fears play an important role. In UNDRIP, IP rights are connected with the concepts of cultural heritage and traditional knowledge, upholding the collective character of these rights. Viewed in this light, "the protection of the intellectual property of indigenous peoples becomes inseparable from the protection of their broadly defined cultural rights . . . [and] necessary to the realization of their fundamental rights to self-determination and to own and control their ancestral lands."[97] And with respect to the 2017 recommendation, the fear of brain drain may have resulted in a wish to use the promise of IP rights to make it as attractive as possible for young researchers to stay in or to come back to their native countries. In addition, both UNDRIP and the 2017 recommendation are more recent instruments than the ICESCR, reflecting the fact that IP has increasingly become a decisive factor for scientific research and its products.

Another instrument that seeks to protect traditional medicinal knowledge is the Nagoya Protocol on Access to Genetic Resources and the Fair and Equitable Sharing of Benefits Arising from Their Utilization to the Convention on Biological Diversity. Ratified in 1993, the UN Convention on Biological Diversity (CBD) is dedicated to promoting the conservation of biological diversity, the sustainable use of its components, and the fair and equitable sharing of benefits arising from genetic resources.[98] The Nagoya Protocol, which is a supplementary agreement to the CBD, was adopted in 2010 and entered into force in 2014. Though the CBD and its Nagoya Protocol are not human rights instruments as such, they have human rights implications in that they seek to protect the rights of owners of genetic resources from illegal biopiracy.

The main concern is ensuring benefit sharing when genetic resources and the traditional knowledge associated with those resources are utilized. Obtaining prior informed consent by the provider country or community and the establishment of mutually agreed terms are two ways in which such benefit sharing can happen. Though not mentioned directly in the Nagoya Protocol, property, especially IP, is central to benefit sharing.[99] And when it comes to the knowledge production on which IP is based, mutuality is the way forward. The early human rights instruments talked about transfer of knowledge, typically from the global north to the global south. Perhaps a better phrase to signal the importance of mutuality is what Sheila Jasanoff has called "co-production" of knowledge.[100]

Other scholars talk, albeit in slightly different contexts, about knowledge in transit, traveling knowledge, and global history.[101] Knowledge circulates and we should neither study communities of knowledge production in isolation nor understand world history as emanating from the West. Knowledge is not given to someone who passively accepts it; in the case of bioprospecting, there is a sharing of knowledge that should be recognized by everyone concerned—and reflected in systems of IP or authors' rights.[102] From a cultural rights perspective, authors' rights are much to be preferred as they are more in line with the notion of "the significant value of science as a common good," referred to in the preamble of the 2017 recommendation, than are private property rights.

Scientific (Mis)conduct

What can scientists who study and work with medicinal plants do to prevent biopiracy and protect traditional knowledge? There are, Thomas Efferth and cowriters suggest, two ways in which scientists can react here and now.[103] The first is to make sure that local communities are included in patents and get a fair share of the earnings from the commercialization of the products involved. Ethically speaking, this is an "indispensable obligation" both for companies claiming to be interested in sustainability and reciprocity and for scientists who work in academia but collaborate with pharmaceutical or other companies on medicinal plants.[104] The second thing that scientists can do, according to Effert, is present their data and findings at conferences and write these up into academic articles so that they cannot lead to any patents: "Published data are not patentable anymore. . . . The same is true for presenting scientific data at conferences. Even unpublished results once presented in oral form (as speeches or lectures) or as posters cannot be patented thereby depleting the risk of a probable biopiracy attack. This is an easy and effective way for all of us to act against biopiracy. Let us publish more on medicinal and aromatic plants!"[105]

Long term, scientists can facilitate international collaboration between universities and research institutions. This should be done in such a way that students from developing countries, from which originate the medicinal and aromatic plants studied and the knowledge that goes into using them, get the chance to work with their peers and renowned researchers at institutions in developed countries. In order to further the scientific advancement on medicinal and aromatic plants, supervisors should also be involved, visiting each other's institutions and cooperating wherever possible.[106] These suggestions and ideas are in parallel with the importance given in international human rights instruments to international assistance and cooperation. For science and scientific research, such international collaboration, as Efferth and his cowriters express it, is "a win-win situation for all participants"; indeed, without it, there would simply be no science of importance.[107]

In pointing to the individual responsibility of scientists and stressing that there are things each scientist can and should do (and not do), Efferth and his cowriters furthermore echo the arguments made by some bioethicists and legal philosophers. I have written, in previous chapters, about the scientific responsibility that comes with or constitutes a part of each scholar's scientific freedom. In Chapter 4, for example, I discussed the issue of scientific self-regulation and also touched on the so-called precautionary principle, an idea pursued in the EU system but also elsewhere. Here, I will take a closer look at some of the guidelines and procedures that have been established to address dishonest ways of doing science and what gave rise to them.[108]

Scientific misconduct first became an issue of interest beyond academia itself in the United States when cases began appearing in the 1980s of fabricated research by respected scientists at some of the country's best universities. These cases were publicly prosecuted and when the media started writing about them, they spurred the interest of the public.[109] In Europe, the same thing happened about ten years later, and on both sides of the Atlantic, these cases gave rise to a demand for public scrutiny of scientific activities and to the establishment of boards and institutions that can oversee the world of science and target any fraud.

In the European context, the Nordic countries and Germany were among the first to deal officially with scientific misconduct. In order to investigate and to initiate preventive measures against reported cases of scientific fraud, committees were established by the national medical research councils in Denmark (1992), Norway (1994), and Sweden (1997), as well as by the Ministry of Education in Finland (1994). In Germany, the German Research Foundation (Deutsche Forschungsgemeinschaft) issued *Proposals for Safeguarding Good Scientific Practice* in 1998; two years later, the Max Planck Society followed suit with its publication *Rules of Good Scientific Practice & Rules of Procedure in Cases of Suspected Misconduct*.[110]

The Nordic committees have consisted of both legal scholars and scientists, and sanctions or punishment have been the responsibility of the institutions at which the scientists under suspicion of misconduct have been active.[111] In addition to disagreements about authorship, most reported cases have concerned uncertainty with respect to who has the rights or duties to use the scientific data in question. The terms of agreements made at the beginning of joint studies have also led to disputes. Guidelines on good practice have subsequently been developed in all countries.[112]

In 2017, a new European Code of Conduct for Research Integrity was developed just as the members of UNESCO agreed on the Recommendation on Science and Scientific Researchers. All these codes and guidelines reflect the way in which scientific misconduct has become a public issue. Scientific dishonesty and lack of transparency and accountability are by no means new phenomena, however. They have always been there, scientists being human beings just like everyone else. Some of the reasons why things have escalated since the 1980s are well known and well rehearsed: scientific research has become more competitive and costly, scientists are under more pressure to produce original research, to apply for funding (especially private funding as public funding is drying out in many countries), and to publish as much and as fast as possible.[113]

But, Roberto Andorno argues, there is more to it. The scientific method, implying an objective and disinterested way of conducting scientific research, is essentially a myth, which leads the public to have unrealistic expectations of science and scientists. In order to avoid expectations that cannot be fulfilled, says Andorno, it would be better "to regard the 'scientific method' as an *ideal* to strive for (even knowing that it is unattainable in its fullest form) and not as the description of an actual practice in scientific research."[114] In her attempt to answer why we should trust science, Naomi Oreskes makes a somewhat similar argument, as we saw in Chapter 1. We should not trust science, she says, because of the scientific method supposedly followed by scientists, but for the reason that the scientific process entails a rigorous vetting of claims and a process of discussing, negotiating, changing, and verifying results. At the end of this vetting process, consensus will build and this is when we can trust scientific knowledge. Oreskes' book came out before the COVID-19 crisis, but her claim that scientists may point to the substantial record of success provided by modern medicines and technologies as evidence that they have done something right is one that more people would today agree with than before the coronavirus came into their lives.[115]

Conclusion

I suggest that we read ICESCR Article 15 from the bottom up, starting with science as an international endeavor [Article 15(4)]. Without inspiration and input from fellow scientists, citizen scientists, and others around the world, individual scientists and their research groups will develop none of those original ideas that lead to progress, short or long term. Without scientific freedom [Article 15(3)], there will be no pursuit of ideas. There will be nothing to disseminate [Article 15(2)]—and therefore nothing for the public to benefit from, participate in, and share [Article 15(1)].

Without a flow of ideas from around the world, that is, there will be no research output from which the public can benefit. The scientific process thrives on inputs made, and subsequently solidly tested, refuted, or validated, by groups and individuals from all corners. It is therefore imperative that neither economic forces utilizing the IP system nor political or ideological concerns relating to that which is native and good versus that which is foreign and questionable succeed in stopping the free flow and exchange of creative ideas.

Like good ideas, plants do not respect borders. In this chapter, we have seen how difficult it can be to establish which plants are native and which plants are foreign. Native or foreign, moreover, some plants thrive, while others become extinct because they cannot adapt. This was the case at the time of the *Flora Danica* project—and it is still the case today. Foreign plants are sometimes labeled invasive or invasive alien plants. Ground elder (*Aegopodium podagraria*), illustrated in *Flora Danica* plate 670 (1777), is one example of such an aggressive or dynamic plant.[116] Its Latin name seems to point to its use in popular medicine against gout, its English name to the similarity of its leaves and flowers to the elderberry plant.

Some contend that monks introduced ground elder from southern Europe during the Middle Ages; others claim that it has been present in the Nordic countries much longer and that we cannot tell where it came from.[117] Today, most people know it from gardens and other places around inhabited areas—mostly as a nuisance weed that is difficult to get rid of in flowerbeds or vegetable patches. On the so-called list of plants under observation, kept by the Danish Ministry of Environment and Food, ground elder is one of twelve plants that are not yet but probably will become invasive.[118]

At the same time, however, ground elder has become a much-used ingredient in New Nordic cuisine.[119] One person's weed here literally is another person's delicacy! Invasive or not, aggressive or just plain dynamic, ground elder has succeeded in adapting to changing circumstances. When Nordic chefs use plants such as ground elder in their culinary experiments, they are motivated by traditional Greenlandic ways of cooking and preparing food.[120] In Greenland, which is today no longer a Danish colony but a part of the Danish realm with home rule, traditional food cultures use seasonal ingredients that are readily available locally. This has inspired the Nordic chefs. But inspiration has also gone in the opposite direction. A market for a so-called new Greenlandic food culture has recently been created in Greenland.[121]

When it happens on mutually agreed terms, this is the kind of inspirational flow of ideas and knowledge sharing that are so crucial for the activation of the right to science (as well as for the right to culture).

CHAPTER 6

Of Human Rights, Human Duties, and Science Diplomacy

Endorsed by the global community in 2015, the UN 2030 Agenda for Sustainable Development is firmly "grounded in the UN Charter, the Universal Declaration of Human Rights, international human rights treaties" and other instruments such as the Declaration on the Right to Development.[1] The seventeen sustainable development goals (SDGs) seek to "realize the human rights of all" and "to leave no one behind."[2] There is no stand-alone goal on science, but inputs from science, broadly understood to include engineering as well as the social sciences and the humanities, are required for all SDGs,[3] scientific knowledge being "in its pure form . . . a global public good."[4]

Many of the SDG targets reflect the moral principle that every country—whether poor, rich, or middle-income—carries a responsibility for realizing the SDGs. SDG 17, Partnerships for the Goals, most directly addresses duties of international cooperation and the right to development, but many other SDGs also hold challenges for both developed and developing countries.

What part each country may play naturally depends on available resources. Countries may also think differently about which SDGs and targets are the most important for them. While some SDGs target the needs and aspirations of developing countries, others specifically draw attention to the responsibilities of the developed world to make appropriate changes, show solidarity, and help where it can.[5] The message is unequivocal: we are in this together, and we will manage to create an economically, socially, and environmentally friendly world only if we all contribute[6]—"Universality implies that all countries will need to change, each with its own approach, but each with a sense of the global common good. . . . Universality embodies a new global partnership for sustainable development in the spirit of the UN Charter."[7]

The COVID-19 pandemic has provided an excellent illustration of the importance of scientific international cooperation. According to ICESCR Article 15(4), as we saw in the previous chapter, international cooperation makes investment in knowledge sharing necessary not only to counteract global pandemics but also to mitigate the risks deriving from the disparities among countries in terms of scientific knowledge and access to its applications. Such international collaboration is not limited to state-to-state interactions. International agencies must also collaborate with each other and with states parties, NGOs, academia, and the general public to ensure the efficient use of their resources. In the context of the COVID-19 pandemic, international agencies such as the WHO, the World Intellectual Property Organization, and WTO have been especially relevant just as international cooperation between public and private entities has become important.

There have been wide disparities among nations with regard to whether they have based their COVID-19 policies on scientific evidence. These disparities track inequalities in health outcomes, with some countries demonstrating early and effective interventions and relatively low case numbers, and other countries deciding on approaches that led to higher infection and mortality rates.[8] While the pandemic has offered examples of collaborative behavior of states and organizations trying to help others, the COVID crisis has also laid bare behavior that is self-serving.[9] Many countries have thought of their own citizens first and have simply refused to share scientific knowledge as well as vaccines, face masks, and other protective gear with other countries. When the pressure is on and decision making is required for agency, it seems, the global solidarity proclaimed in the international human rights instruments at times fails to materialize in practice.[10]

In this respect, the situation during the pandemic is but a recent illustration of a problem with the international human rights system whose philosophical roots are much older than COVID-19, namely the preference for rights over duties. "It is time to talk about human responsibilities," members of the InterAction Council suggested when they proposed a Universal Declaration of Human Responsibilities in 1997, in preparation for the fiftieth anniversary of the UDHR. "Because rights and duties are inextricably linked, the idea of a human right only makes sense if we acknowledge the duty of all people to respect it. Regardless of a particular society's values, human relations are universally based on the existence of both rights and duties."[11] Coming at the issue of duties versus rights from another perspective, philosopher Onora O'Neill asks in her recent work whether justice and rights without ethics is a twentieth-century innovation.[12]

In the second of its three parts, this chapter discusses the criticism of the international human rights system for not confronting the issue of duties. The first part finishes the story of the *Flora Danica*. It looks at what happened to the work between 1771, when Georg Christian Oeder stepped down as editor, and 1883, when Johan Lange brought it to a successful conclusion, and briefly assesses the *Flora Danica* effort as an early example of both citizen science and science diplomacy. The latter is also the focus of the third and final part of the chapter. Though it may never have been more important, science diplomacy—the use of scientific, technological, and academic collaborations among nations and regions to address common problems and to build constructive global partnerships—is not new.[13] During the twentieth century, scientists became increasingly involved in political and diplomatic matters, and today, "with government co-operation at an all-time low and America pulling funding and membership from the World Health Organisation," writes Nobel Prize–winning biochemist Jennifer Doudna, "we must rely on a renewed push for 'science diplomacy.'"[14]

Science diplomacy, I argue, illustrates the importance of international collaboration recognized in ICESCR Article 15(4). It also highlights the necessity of scientific freedom under responsibility, outlined in Article 15(3). Without scientific freedom, the safe exchange of scientists, scientific knowledge, and technological equipment cannot be guaranteed and no scientific progress can be made from which the public may benefit.[15]

Flora Danica: Finishing the Story

A year hence the classical Flora Danica will be terminated
by the completion of the seventeenth volume. The
work will contain figures of 4,000 species of plants, of
Scandinavia, including Greenland and Iceland. It has been
published wholly at the expense of the King of Denmark,
and a right royal work indeed. At its completion the
plates (in folio) which relate to Greenland plants, and
which illustrate its whole flora, are to be separately
issued, with a brief letter press, under the title of Icones

> Florae Groenlandicae. As this flora is in one sense
> American, and as the copies of the whole Flora Danica
> in the United States are and must be very few, we take
> pleasure in announcing this illustrated Greenland Flora
> to American botanists. Some of them will wish to possess
> it. The price of uncolored copies is fixed at 56 francs, of
> the colored at 236 francs. It should be added that, as
> the impression is strictly limited, application should be
> made very promptly. The editor, Professor Joh. Lange,
> Copenhagen, informs us that he will himself receive
> subscriptions, up to the first of May next.[16]

Signed "A. G.," this entry is from the March 1882 issue of the *Botanical Gazette*, a journal of plant sciences published by the University of Chicago Press now known as the *International Journal of Plant Sciences*.

"A. G." was probably Asa Gray (1810–1888), who is today acknowledged as "the father of American botany."[17] Gray was elected to the American Academy of Arts and Sciences in 1841 and accepted a professorship of natural history at Harvard College the following year, at age 32. He donated his private collection of books and plants to Harvard in 1865. Within a few years, what had begun as a private library had grown considerably and now contained full sets of valuable periodicals and expensive works. The *Flora Danica* was one of these.[18]

Gray was interested in the resemblance of American flora to those of Europe and Asia and wrote in an early paper on the distribution of plants that "it seems almost certain that the interchange of alpine species between us and Europe must have taken place in the direction of Newfoundland, Labrador, and Greenland, rather than through the polar regions."[19] This may be the reason why he wanted to help promote the illustrated Greenlandic flora.[20] As mentioned in Chapter 5, *Icones Florae Groenlandicae*, with 336 plates of illustrations of Greenlandic plants, was published in 1883, the same year the last issue of the *Flora Danica* itself appeared. The editor of both was Johan Lange.

Between the editorships of Oeder and Lange, a number of other botanists were involved in editing and publishing the *Flora Danica* for shorter or longer periods of time. All were good botanists and scholars, but some were more dedicated to and had a greater impact on the work than others. Otto Friedrich Müller (1730–1784; editor of *Flora Danica* 1775–1782), who took over from Oeder, was one of the less dedicated editors. A self-taught botanist and zoologist, Müller was quite well known in Europe as the publisher of a volume on the flora of plants to be found in the part of Denmark where he lived. He eventually developed more of an interest in zoology, especially minute animals, and his *Zoologica Danica*, intended to correspond to the *Flora Danica*, earned him more fame than did the volumes of the latter that he succeeded in editing. These are, in one estimate, "the poorest in the work. [Müller's] identification are not precise and the drawings are often poor."[21] Müller did, though, add to the work a variety of fungi and algae, especially microscopic algae, to which Oeder had not paid much attention.[22]

Norwegian-born Martin Vahl (1749–1804; editor of *Flora Danica* 1785–1799) succeeded Müller. Much like his predecessor, Vahl was involved in other scholarly endeavors that took time and energy away from his editing of *Flora Danica*. In his other publications, he described hundreds of new plant species, but he added very few species to the *Flora*, with Norwegian plants dominating.[23] One of his other major publications was a book describing plants he had received from the Danish West Indies, as mentioned in Chapter 4.[24] Another was *Symbolae Botanicae, Or, Exact Descriptions of Plants Collected by Petrus Forsskål on His Arabian Journey*, published between 1790 and 1794. A student of Linnaeus, Forsskål (1732–1763) participated as the natural historian in one of Europe's first truly scientific expeditions to the Arabian Peninsula, sent out in 1761 by Danish king Frederik V (1723–1766). Forsskål was the first person to describe the unusual plant and animal life of the Red Sea, but sadly he died at age thirty-one in Yemen in 1763. The German surveyor Carsten Niebuhr was the only member of the expedition to survive. He saved the greater part of the expedition's collection and brought it back to Denmark where Vahl took it upon himself to work through Forsskål's botanical material.[25]

Like Forsskål, Vahl studied with Linnaeus. He spent the years between 1769 and 1774 in Uppsala—years that, together with his extensive travels around Europe, where he was given access to some of the best and most comprehensive herbaria, convinced him that the Linnaean system "was buckling under the weight of impenetrable taxonomic mistakes."[26] He conceived the ambitious plan of improving Linnaeus's system by producing a catalogue of all plants in the world, including the plants from faraway places such as the West Indies and the Arabian Peninsula that were increasingly made available to European botanists at this time. He worked on the *Enumeratio Plantarum*, as he called this catalogue, during the last years of his life, but succeeded in publishing only the first volume shortly before he died in 1804. The second volume was published by some of his friends the following year, but the rest of the about twenty planned volumes were never completed.[27] Vahl left about twenty-six thousand handwritten index cards in Latin, hinting at the intended contents of the rest of *Enumeratio Plantarum*. Many of these have been digitized and are still consulted regularly by students and scholars of botany.[28]

A student of Vahl, Jens Wilken Hornemann (1770–1841) was the editor of *Flora Danica* for thirty-five years, from 1805 to 1840. He traveled to all parts of Denmark and the dukedoms, and his dedication to *Flora Danica* was more consistent than that of his predecessors. We have already come across this longest-serving editor in various connections in previous chapters. It was Hornemann who was one of the first Danish botanists to attempt to evaluate whether plants to be found in Denmark were native (see Chapter 5). It was also Hornemann who first started including Greenlandic plants. This seems to have happened to compensate for the lack of access to Norwegian plants after 1814 when the Danish king was forced to cede Norway (though not the old Norwegian dependencies of Iceland, the Faroe Islands, and Greenland) to the king of Sweden after having been on the losing side during the Napoleonic wars (1803–1815).[29]

During his travels around Denmark and the dukedoms, Hornemann noticed that there was a problem with the distribution of *Flora Danica* copies. As mentioned in Chapter 1, the original wish had been to engage voluntary collaborators, especially gentleman farmers and clergymen, but also students and correspondents, throughout Denmark. Copies of the work would be made available to bishops and chief administrative officers so that whoever wanted to could then borrow them and in return share their own observations with Oeder and his colleagues in Copenhagen.[30] When Hornemann discovered that a number of these copies either were incomplete or had never reached their destination, he made sure defective copies were repaired and that all copies were moved to the diocesan libraries.[31] He also made it a priority to look during his travels for interested lay people or, as we would say today, citizen scientists. In this way, he helped copies finally find their way to botanists and others, "many of whom were to make important contributions to the work."[32]

After Hornemann died in 1841 and until Johan Lange took over as editor in 1858, four different botanists were involved with *Floral Danica*. As two of them, Joachim Frederik Schouw (1789–1852) and Frederik Michael Liebmann (1813–1856), had studied with Hornemann and a third, Jens Vahl, was the son of Martin Vahl, it is fair to say that the editorship more or less stayed in the family.[33] The fourth botanist was Salomon Drejer (1813–1842), a productive and promising young scholar who unfortunately died young and contributed to only a single fascicle. Schouw was also educated as a lawyer and was to play a role in Danish politics, actively canvassing for change of the political system and the abolition of absolutism.[34] He was an important initiator of and participant in the meetings of Scandinavian natural scientists that took place during the 1840s. At the meeting in 1847, it was decided to include the Norwegian and Swedish plants that had not already been illustrated. The result was three installments or supplements published as a *Supplementum Florae Danicae.*

Liebmann, who had successfully applied for the position as editor of *Flora Danica* when Schouw died, published the first of these three supplements. Like Drejer, however, Liebmann sadly died young. Before his brief period as *Flora Danica* editor, he had traveled to Mexico, had visited the Pacific Coast, and had sailed back to Denmark via Cuba, everywhere collecting plants and living specimens. His was "one of the most successful expeditions ever made by a Danish botanist," and some of his specimens are still alive at the Botanical Garden in Copenhagen.[35] Though performing an important job as assistant to Schouw and being responsible for some *Flora Danica* plates, Jens Vahl was not considered qualified enough to be appointed editor in his own right.[36]

Johan Lange published the other two *Supplementum Florae Danicae.*[37] Introducing the third and last of these at a meeting on March 27, 1874, of the Royal Danish Academy of Sciences and Letters, he added the following comments:

> As members of the Academy will know, a royal rescript, issued on 9 October 1847, prescribed an expansion of the work 'Flora danica' by the addition of Swedish-Norwegian plants that do not grow naturally in Denmark or that have not earlier been illustrated in the work, which until 1814 also included Norway's flora. . . The first supplement, which implemented this lucky and in terms of plant geography very natural expansion of the work to include all of Scandinavia as well as also the northern dependencies (Greenland, Iceland, the Faroe Islands), was published by the late Professor Liebmann and announced by him at a meeting in the Royal Academy on 1 April 1853.[38]

With his long tenure as editor of *Flora Danica*, from 1858 to 1883, Lange provided much-needed stability. On April 30, 1880, at another meeting at the Royal Danish Academy of Sciences and Letters, the last editor of *Flora*

Danica was finally able to announce that "the illustrated work will be concluded in the spring of 1883, where after the whole considerable work, which
was started 120 years ago (1761), will be concluded with a register and a critical revision of its contents."[39]

Flora Danica: A Work of Citizen Science and Science Diplomacy?

A great number of people were involved in the *Flora Danica* endeavor. In
addition to the editors themselves, there were the printers, the engravers, and
the illustrators. Their salaries were all paid by the Danish kings, who also
sponsored the various expeditions to collect relevant plants. And then there
were what Henning Knudsen, in his magisterial 2014 work on the *Flora*, calls
"the florists." Some of these were trained botanists; others were nonspecialists
interested in finding and investigating plants in nature. "They all contributed
to the final quality of *Flora Danica*. . . . Some delivered sketches for drawings,
others approved drawings, some delivered descriptions, and still others made
their material available for the publishers."[40]

When Hornemann and other editors investigated whether copies of the
work actually reached their intended audience, they were worried about losing the feedback and contributions from precisely these "florists" or citizen
scientists who were so important to the creation of *Flora Danica*.[41] An 1842
revised plan for the work shows how integral this exchange of copies for
information was to the production process: "With regard to the distribution
of copies, the rule shall be this, that a copy of the engraved work Flora Danica
shall be kept in every diocesan library for consultation and use on the spot,
as well as another, which may be lent to individual botanists in the diocese,
who may then make contact with the publisher and shall be obliged to report
to him such information as may be useful to the publication of the work."[42]

Is it fair to describe *Flora Danica* not only as an early example of citizen
science but also as a tool of science diplomacy? I think it is. We have heard in
previous chapters how the work not only was praised by foreign botanists but
also set an example for other European floras, and how it was used by succes-
sive Danish kings for diplomatic reasons. In one early assessment, the work
was seen as an illumination for all of science.[43] Almost eighty years later, Asa
Gray called *Flora Danica* "a right royal work indeed." It was published in several languages, so that it would attract both a domestic and an international

audience. Expensive as it was, complete copies were given as a gift to only a few famous botanists such as Linnaeus in Sweden and Joseph Banks in England, as well as to noblemen and members of royal families deemed worth impressing. Copies presented as gifts would typically end up in institutions; for example, Joseph Bank's copy was given to the British Museum.[44] Today, complete sets are rare. Notes Henning Knudsen, "Outside Denmark, 13 copies of *Flora Danica* are known in the USA, six in the UK, three in Austria, with others in libraries in Helsinki, Haarlem, Firenze, Genève, St. Petersburg, Madrid and Paris."[45]

In the domestic context, the work did not quite live up to the wish of the Danish kings for a practical instrument which could enhance the use of local plant resources and thereby benefit the country's agricultural economy. Those texts of explanation and description, originally promised by Oeder, never appeared, and the whole process of editing and publishing the work simply took too long. In some ways, the aim of using *Flora Danica* for science diplomacy clashed with the intention of providing a tool for citizen scientists who could use it when and how they pleased. In his 1761 invitation to subscribe to "a new work called *Flora Danica*," Oeder suggested that subscribers "consider the collection of engravings as a so-called *Herbarium vivum*, which one may collect and increase when there is an opportunity for it, and which one may organise in an appropriate order when a sufficient number of engravings has been collected."[46] Oeder's superiors did not approve of such an individually targeted or customized approach, however. As they saw it, "a collection like that would lose all the prestige of being a Regal work, and a work which by its magnificence and costliness was a credit to the nation."[47]

These clashing objectives notwithstanding, in combination with the *Flora Danica* dinner set, the work has become an iconic piece of Danish culture that keeps inspiring artists, designers, and other cultural workers. And that, as Knudsen puts it, "is not too bad for a scientific piece of work with something as highbrow as a Latin title."[48]

Human Rights and Human Duties

Just as natural historians of the eighteenth century such as Oeder and his colleagues attempted to bring order to the known living world by developing a universal and workable methodology of classification of plants (and animals), so in jurisprudence "natural law" was the term used for any kind of regularity.

The 1776 American Declaration of Independence invoked the "unalienable Rights" to which the "Laws of Nature" and "Nature's God" entitle people.[49] The 1789 French Declaration of the Rights of Man and of the Citizen also appealed to the authority of the natural order. Both were "the faint echoes of a shared history" between natural history and law.[50] Eventually, both natural historians and jurists would become more hesitant to appeal to God as the direct source of law. The former would increasingly rely on scientific experimentation, empirical data, and rational reasoning just as jurists would become more and more interested in positive and man-made law rather than in divine law.

The conception of natural law as meaning or implying natural rights has continued to play a role in notions of rights in domestic and international contexts, however. Today, it has largely been replaced by the expression "human rights."[51] Scholars still discuss when that expression was vernacularized and who the major protagonists were, but most tend to agree that those late eighteenth-century American and French declarations of "unalienable" and "natural rights" are important historical antecedents.[52] The doctrine of natural rights came under immediate philosophical and political fire; Jeremy Bentham famously called rights "nonsense upon stilts," for example.[53] Though by now playing a solid role in political debate and dominating the texts of legal declarations and treaties, those rights and their definition and legitimacy are still the subject of heavy discussion: "Whether human rights are to be viewed as divine, moral, or legal entitlements; whether they are to be validated by intuition, custom, social contract theory, principles of distributive justice, or as prerequisites for happiness; whether they are to be understood as irrevocable or partially revocable; whether they are to be broad or limited in number and content – these and kindred issues are matters of ongoing debate and likely will remain so as long as there exist contending approaches to public order and scarcities among resources."[54]

In a 2018 article, Malcolm Langford analyzed a number of social science–inflected critiques of human rights.[55] These critiques fall into three broad thematic categories: sociological legitimacy, material effectiveness, and distributive equality. The sociological legitimacy critique concerns the claims for universal human rights. These rights are, maintain postcolonial and post-structuralist critics especially, Western in origin and not only have actively promoted Western individualism at the cost of the communitarian views and group rights valued in other parts of the world; they have also encouraged Western imperialism.[56] We have already met this critique in previous chapters, especially Chapter 5.

The second category, the effectiveness critique, comprises a wide range of warnings about the futility of human rights standards, advocacy, and policy. Scholars in fields ranging from sociology and political science to economics, anthropology, and social psychology argue that human rights have never been properly enforced but have remained just pretty words on paper. States have delayed or simply resisted carrying out decisions made by international courts, just as rights-based efforts by civil society organizations have not had much of an impact, the overall effect being that not much progress has been made on key human rights.[57]

When it comes to the redistribution of democratic power as well as of wealth and resources, argue scholars grouped into the last of Langford's three categories, the distributive equality critique, the human rights egalitarian project has been especially inadequate.[58] For all the talk in official UN documents of democratic access and participation, precious little restructuring of power relationships has taken place in practice. We have also not seen a fairer distribution of economic resources, scholars maintain, drawing attention to the poor record of both legal and political activism in this area.

To Langford's three thematic categories of human rights critique, I would like to add a fourth: the duties and responsibilities critique. Sharing the worry expressed by both the material effectiveness and the distributive equality critique, scholars within this category argue that there is today too much talk of rights and legal obligations and too little talk of individual and non–state actor duties and responsibilities. This is a challenge for policy making, but it also presents a problem from an ethical, and some would say a religious, point of view. "Obligation" is the word used in human rights instruments about duties that are correlative to the described rights. For example, the right to vote imposes a corresponding duty on the state to make sure that elections take place. In terms of policymaking, when states fail to meet their correlative duties or to implement rights, other relevant agents, including individuals and collective groups, must be identified who can take responsibility for bringing about political and normative change, writes international relations scholar Kathryn Sikkink.[59] She argues for a "rights-and-responsibilities framework" that could identify relevant agents who may work together toward realizing rights in practice.[60]

Ethically speaking, contends philosopher Onora O'Neill, one problem is that the major human rights instruments are silent about duties without counterpart or correlative rights and that the explicit language of individual duty to other individuals, to society, or to the state is reduced to a matter of subjective, private values without any deeper justification.[61] Individual or collective cultural and subjective preferences or values are not the same as ethical claims about how to act fairly. "Standards of honesty and reliability, standards of confidentiality and fairness, standards of trustworthiness and discretion, and many other ethical standards are not irrelevant to justice, and cannot be replaced by ever more detailed elaboration of law, regulation and guidance."[62] For centuries, "European traditions had seen [rights and duties] as contributing distinct but parallel answers to the classical question 'what ought we to do?'"[63] Now, neither legal nor political philosophers like to talk about ethical justifications for principles of justice, but seem to be interested only in that which complies with the letter of the law.

The terms "duty" and "responsibility" can also be associated with religion. Among the council members and supporters of the initiative to draft the 1997 Universal Declaration of Human Responsibilities were both political and religious leaders, for example.[64] In the introductory comments, the initiative is described as "not only a way of balancing freedom with responsibility, but also a means of reconciling ideologies, beliefs and political views that were deemed antagonistic in the past."[65] Several former African political leaders endorsed the declaration. They most likely brought to the table ideas and inspiration from the African (Banjul) Charter on Human and Peoples' Rights, adopted in 1981, which not only insists on duties alongside rights but also stresses the importance of the collective rights of peoples in addition to individual human rights.[66] The declaration never had much of an impact, but it constitutes a significant attempt to bring together religious and nonreligious beliefs and efforts and to reconcile communitarian African and Asian values with more western notions of universal rights.[67]

Countervailing or Redemptive Visions

Together, the various and at times overlapping points of criticism raised by scholars in the past many years clearly show, in the words of Malcolm Langford, that "the slow march of rights should be met with a critical reflex."[68] Even so, there are countervailing, even redemptive narratives that are worth our attention. One of these, it seems to me, is the 2030 Agenda for Sustainable Development, which reflects a willingness to meet several of the objections raised against the human rights regime. The agenda differs from previous UN initiatives in at least three ways. It makes the deep, global societal problems stemming from an unacceptable imbalance of power and wealth everybody's business and responsibility; it points to ways of implementing key rights; and it has science and technology play a key role. The ethical principle behind the agenda is a sense of universal responsibility for the state of our world that is felt to be missing by critics in both the sociological legitimacy and the duties and responsibilities camps. In addition to helping countries that cannot on their own live up to the goals of the agenda, the developed world has to make significant domestic changes in order to fulfill the pledge of leaving no one behind that guides the implementation of the SDGs.

The SDGs are not directly phrased as human rights, but several of their 169 associated targets are closely linked with human rights standards.[69] Of special significance in our context is the fact that a number of these fall into the category of economic, social, and cultural rights. "The 2030 Agenda for Sustainable Development is the renewed common global commitment of States to eradicate poverty in all its forms and dimensions," as the CESCR puts it in a 2019 statement.[70] For the effort toward achieving distributive equality, the ICESCR is a key instrument, and human rights scholars would benefit from engaging more with economic, social, and cultural rights, in my opinion.

As the CESCR states, the SDGs are based on the very rights that are protected under the ICESCR. These include, for example, the economic and social rights to work (Goal 8), to social security (Goals 1, 2, 3, 5, and 10), and to an adequate standard of living (Goals 1 and 2, 6 and 7, and 11 to 16). They also comprise cultural rights such as the right to take part in cultural life (Goal 16 and relevant targets in other goals), the right to education (Goal 4), and the right to enjoy the benefits of scientific progress and its applications (Goals 9 and 10). Goal 10 obligates states to reduce inequality within and among countries.[71] Finally, Goal 17 is intended to reinforce the means of implementation:

> the 2030 Agenda highlights the imperative to strengthen domestic resource mobilization by improving domestic capacity for taxation and other revenue collections. At the same time, domestic resource mobilization can and should be supported through international cooperation and assistance to developing countries through official development assistance, and the use of other resources. . . International cooperation in this context includes not only financial resources, but also access to relevant technology needed for sustainable development and capacity-building.

Goal 17 further highlights that States should not harm developing countries by preventing them from making the necessary policy choices, for example, in the sphere of trade that could impact negatively on those countries' ability to fully implement the Sustainable Development Goals.[72]

That is, when it comes to the implementation of the SDGs and their targets, science and technology form an important part of that international cooperation without which the full realization of economic, social, and cultural rights is not possible. "For the first time at this level," wrote Irina Bokova, then director-general of UNESCO, in her foreword to the 2015 *UNESCO Science Report: Towards 2030*, "the role of science, technology and innovation has been explicitly recognized as a vital driver of sustainability."[73]

Bokova made her statement before the COVID-19 pandemic hit in 2020. If anything, as discussed in Chapter 5, the pandemic has made the importance of scientific research and output—and the right to science—even more obvious. As one of those economic, social, and cultural rights that the drafters of the UDHR considered to be "indispensable for [a person's] dignity and the free development of [their] personality,"[74] the right to science, along with the other rights listed in UDHR Articles 23 through 27, is groundbreaking because it aims at the realization of the development of one's self.[75]

With the four parts of ICESCR Article 15, the drafters of the ICESCR continued this line of thinking—a line of thinking that is also reflected in official UN documents and statements, especially recent ones, as I have attempted to show in this book. The very fact that the right to science was originally categorized as a cultural right [Article 15(1)] enables us to see science, done by both professional and citizen scientists, as a part of culture and to apply ethical and cultural concerns to scientific scholarship just as we do to any other kind of scholarship. It also allows us to approach the issue of authors' rights [Article 15(1)(c)] and IP in such a way that legal discourses no longer automatically protect IP over human rights and social values.[76] As I read the last part of the quotation above from the 2019 CESCR statement, this approach is relevant to the issue of trade.

In order to enable dissemination [Article 15(2)], scientific freedom [Article 15(3)], and global cooperation [Article 15(4)], open access to scientific results is crucial.[77] This is highlighted in all relevant UN documents as well as in the 2030 agenda, which understand science, both applied and basic,[78] as "a way of crossing national, cultural and mental borders."[79] For the agenda to succeed, moreover, scientists themselves need to take more responsibility and play a more active role—or at least to realize that it is not enough for them to produce scientific analyses and then leave the task of acting on those analyses to policymakers (alternate side of scientific freedom).[80] Geopolitical developments such as the current trade frictions and technological contests between the United States and China impede the global free flow of people and ideas and open access to scientific results, without which there will be no scientific progress from which the public and society can benefit. Thus it is necessary for scientists at times to engage in science diplomacy—a kind of activity for which they are not trained and therefore often feel unprepared.

Science Diplomacy

Science diplomacy has become an umbrella term for the use of scientific and academic collaborations among nations to address common problems. The idea is to use science as a "soft" or "smart power" tool to achieve foreign policy objectives. During the Cold War, not only research outcomes but also science itself as a process and way of communicating were used to further specific power interests. But science diplomacy may also serve to build constructive international partnerships. One example is the Synchrotron-Light for Experimental Science and Applications in the Middle East laboratory (SESAME), one of the most high-profile science diplomacy initiatives of the twenty-first century, which opened in Jordan in 2017. Like the European research center CERN, SESAME was created after a UNESCO decision and is today a fully independent intergovernmental organization.[81]

The AAAS has its own Center for Science Diplomacy as well as its own journal, *Science & Diplomacy*.[82] The British Royal Society has worked with the AAAS on the scientific dimensions of defining challenges of the twenty-first century such as climate change, food security, poverty reduction, and nuclear disarmament.[83] The EU wants to play an increasingly active and visible role in international science diplomacy "by using the universal language of science to maintain open channels of communication in the absence of other viable foreign policy approaches,"[84] and in Germany, the government wants to turn the Federal Foreign Office into a laboratory for science diplomacy.[85]

Research and academic relations policy has always been an important part of German foreign policy, but a new strategy is needed for the 2020s. This is the message of a 2020 German foreign policy strategy paper. The COVID-19 pandemic and the way in which it has been dealt with in Europe and elsewhere clearly shows, this paper states, that "science can provide the fact-based foundations for political decisions."[86] For this reason and, more generally, to increase Germany's attractiveness as a place to study and pursue research, the German Bundestag and relevant ministries such as the Federal Ministry of Education and Research have issued a set of guidelines in order to implement the new strategy. From now on, German science diplomacy, which is squarely embedded in European science diplomacy, will work toward responding to "the central challenges of the 21st century" by creating and protecting "a space for academic endeavor," and by "promot[ing] and safeguard[ing] academic research as a necessary condition for democratic action."[87]

Science diplomacy is not new. Depending on the definition of both "science" and "science diplomacy," it dates back to at least the mid-nineteenth century, or even to the early modern age. Recognizing science as a global and collaborative undertaking, the Royal Society, founded in 1660, instituted the post of assistant to the secretaries for foreign correspondence in 1723, for example, nearly sixty years before the British Government appointed its first foreign secretary. This post was "the modest antecedent of the Society's present-day Foreign Secretary."[88] The first holder was Philip Henry Zollman, whose role was to conduct correspondence with overseas scholars in order to keep the society's fellows up to date with what was happening in the world of science.[89]

During the twentieth century, scientists became increasingly involved in political and diplomatic matters. In the early 1940s, British scientists Julian Huxley and Joseph Needham played an active role in science becoming a part of UNESCO's title and goals.[90] Huxley became the first director-general of UNESCO and Needham the first head of the natural sciences unit in December 1946. Nine years later, Albert Einstein endorsed with Bertrand Russell and nine other prominent scientists a manifesto calling on scientists to "assemble in conference to appraise the perils that have arisen as a result of the development of weapons of mass destruction."[91] The first meeting of the Pugwash Conference on Science and World Affairs took place in July 1957. Drawing its inspiration from the Russell–Einstein Manifesto, the conference brought together nongovernmental scientists, decisionmakers, and experts to renounce nuclear weapons and find peaceful solutions to threats posed by nuclear and other weapons of mass destruction. Pugwash and its cofounder, Sir Joseph Rotblat, were awarded the Nobel Peace Prize in 1995.[92]

Another prominent and outspoken twentieth-century scientist to realize that the tools, techniques, and tactics of foreign policy need to adapt to a world of increasing scientific and technical complexity was the Danish physicist and Nobel Prize winner Niels Bohr. In October 1945, mere months after atomic bombs were dropped on Hiroshima and Nagasaki, Bohr wrote to his former student, German physicist Werner Heisenberg, that he "firmly believe[d] that the immense implications of the advance of science, by the warning of the necessity of peaceful co-operation between nations, will greatly promote harmonious international relationships."[93] Already during the last two years of World War II, Bohr had been involved in nuclear diplomacy and had met, among others, Roosevelt and Churchill in person. Convinced, like Russell, Einstein, and other fellow scientists, that scientists could help with their unique experience of genuine international cooperation, Bohr emphasized the urgency of making contact with the Soviet Union before the atomic bomb was dropped.[94]

In June 1950, Bohr wrote an open letter to the UN that outlined his views on nuclear weapons and his efforts during and after the Second World War to convince world leaders to be more open about these weapons. "The aim," he wrote, "of the present account and considerations is to point to the unique opportunities for furthering understanding and co-operation between nations which have been created by the revolution of human resources brought about by the advance of science, and to stress that despite previous disappointments these opportunities still remain and that all hopes and all efforts must be centered on their realization."[95] The Korean War broke out only a couple of weeks later, however, and this drew the attention of members of the UN and political leaders away from Bohr's letter. The letter therefore never had much of a political effect at the international level, but it did inspire Scandinavian leaders and governments and would serve as the basis for their approach to nuclear disarmament for many years.[96]

Today, the protagonists are different. But just as science and technology played a major role during the Cold War, so they are at the heart of the current tension in trade relations between the United States and China: "The vast majority of the issue is tech, not trade. . . . The 1960s brought us the global space race. We are now in the next state of the global data race. . . . Those countries that are able to harness data from both an offensive and a defensive perspective for economic prosperity and national security, will be ahead of others."[97] Much like then, countries around the world now face awkward choices between the different political and economic worldviews and divergent global interests of two major powers pulling in opposite directions.[98]

Science Diplomacy Now—and for the Future

As the United States, under President Donald Trump's "America First" policy, began to draw away from the international stage, China became a stronger presence in various UN bodies. At the end of 2018, the United States left UNESCO, which it cofounded after World War II to foster peace, and announced in May 2020, in the middle of the coronavirus pandemic, that it would be terminating its relationship with the WHO.[99] The Chinese effort to expand its global influence and to fill the gap left behind by the United States relies heavily on science and technology. One example is the Belt and Road Initiative, a vast collection of development and investment initiatives stretching from East Asia to Europe. Sometimes referred to as the New Silk Road, this initiative, while intended to usher in an era of political and economic influence for China, also has "established a significant scientific and technological component, including its own international scientific organization."[100]

Another example is China's heavy investment in building advanced research networks, many of which also stretch far beyond China itself. Thousands of Chinese students and scholars have studied in the United States or Europe, and with the exception of U.S. scientists, Chinese scientists today publish the highest number of scientific papers in the world. An increasing number of those papers are the result of coauthorship, most often with U.S.-based scientists (more than 40 percent, according to one estimate).[101] Until now, this widespread scientific collaboration between the two major powers has mostly given rise to concerns about scientific espionage and theft of advanced research and technology, violating IP rights and perhaps leading to dual use technology for military purposes. But it does offer some hope of overcoming political differences in order to work toward common, science-based solutions in the future:

> Recall the lessons from the Cold War. One is the need to focus on areas and topics of mutual interest and concern, such as space, cutting-edge energy projects, and global health. Another is to focus on building institutional links, either by taking advantage of existing institutions of science or, when opportunities arise, creating new ones. In this endeavor, nongovernmental or quasigovernmental organizations are particularly important. But shared interest between the Americans and Soviets around technically based global

challenges such as Antarctica and the loss of the ozone layer also provided an important means to overcome political mistrust to work toward common, science-based solutions.[102]

In 2018, China unveiled a new vision for opening a so-called Polar Silk Road by extending the Belt and Road Initiative to the Arctic region. In its first official Arctic policy white paper, the Chinese government proclaimed that it will encourage the building of infrastructure, paving the way for the development of shipping lanes opened up by global warming.[103] Not itself an Arctic state, China has become increasingly interested in the polar region and became an observer member of the Arctic Council in 2013. The Arctic Council was formally established in 1996 with the Ottawa Declaration, which defined as members of the council the eight Arctic states: Canada, Denmark, Finland, Iceland, Norway, the Russian Federation, Sweden, and the United States. Permanent participant organizations representing Arctic Indigenous peoples also sit on the council, and observer status is granted to non-Arctic states as well as to intergovernmental, interparliamentary, global, regional, and non-governmental organizations.[104]

In a joint communique accompanying the official establishment of the Arctic Council, the ministers of the Arctic states specifically "recognized the contribution of international science to the knowledge and understanding of the Arctic region and noted the role that scientific cooperation, through the International Arctic Science Committee and other organizations, is playing in developing a truly circumpolar cooperation."[105] The International Arctic Science Committee (IASC) referred to here predates the Arctic Council by six years. Representatives of national scientific organizations of the eight Arctic countries established it in 1990, and its founding articles recognize "the importance of the Arctic in advancing world science."[106] Of special interest in our context is the way in which the IASC attempts "to bridge rather than define" human and environmental boundaries and is "committed to recognizing that Traditional Knowledge, Indigenous Knowledge, and 'Western' scientific knowledge are coequal and complementary knowledge systems, all of which can and should inform the work of IASC."[107]

In addition to being one of thirteen non-Arctic states with observer status, China is a member of the IASC. As expressed in the 2018 Arctic white paper, China sees itself as "an important stakeholder in Arctic affairs." Geographically speaking, it is a near-Arctic state and "one of the continental States that are closest to the Arctic Circle," per the white paper. The natural conditions of the Arctic and their changes therefore "have a direct impact on China's climate system and ecological environment, and, in turn, on its economic interests in agriculture, forestry, fishery, marine industry and other sectors."[108] The Chinese white paper reflects the fact that the Arctic has become an important arena for big-power politics with global implications. The main driver in developments, scientific as well as political, is climate change, and the time has come, some argue, to undertake a revision of the international legal rules of governance for the Arctic.

At the 2020 Arctic Frontiers conference in Tromsø, Norway, international relations scholar Bobo Lo suggested that to prevent the Arctic from becoming a competitive arena for territorial rights that may have fatal consequences not just for the region itself but for the entire planet, we need an Arctic treaty. Such a treaty would, he said, include provisions concerning science and the environment as well as politics and security.[109] This is an idea that has come up before and toward which opinions differ markedly.[110] Some have argued that such an Arctic treaty could be modeled on the Antarctic Treaty System, which was created in 1959 and is one of the Cold War triumphs for science diplomacy.[111] It specifies, for example, that "Antarctica shall be used for peaceful purposes only" and that "scientific observations and results from Antarctica shall be exchanged and made freely available."[112]

Others maintain that an Arctic treaty may be a good idea, but that such a treaty should not replicate the Antarctic Treaty. The two poles cannot be compared; whereas no one lives in Antarctica, the Arctic is the home of various indigenous groups, for example. Antarctica furthermore has defined land and sea areas. This is not the case in the Arctic, whose borders to the rest of the world are not agreed on and legally defined.[113] In addition, there are the eight Arctic states that act to protect and maximize their own national interests.[114] At the 2020 Tromsø conference, this became obvious when the Norwegian foreign minister underlined in her comments that the Arctic is well managed and that "we cannot give up on the structures of cooperation that are, actually, working today . . . if something is working, please do not try to revise it."[115]

The Arctic is far from the only political trouble spot in the world at the moment. But several international trends do come together at the North Pole that seem to indicate that a new Cold War could unfold there.[116] As much as the various stakeholders disagree, however, they do see the need to work together to solve challenges they recognize to be global in scope, chiefly climate change. Having such a shared interest is the most important prerequisite for overcoming political mistrust and for working toward common, science-based solutions. This is one of the lessons we have learned from the science diplomacy conducted during the Cold War, as we saw. The stakes are very high and it will take all the diplomatic skills scientists, in cooperation with scientific institutions, NGOs, indigenous groups, and other interested parties, can muster to get a treaty or to provide some other kind of safety net ensuring that the northernmost region of the earth will be used for peaceful purposes only.

ICESCR Article 15 provides support for this important endeavor and guidelines for conduct. Article 15(4) emphasizes the importance of international scientific cooperation and the free flow of people and ideas. These are necessary for scientists to enjoy their scientific freedom [Article 15(3)] and to produce groundbreaking research that can be disseminated [Article 15(2)] for the benefit of all [Article 15(1)]. As interpreted and substantiated by recent UN instruments such as the 2017 UNESCO Recommendation on Science and Scientific Researchers, the 2020 General Comment on Science, and the 2030 Agenda on Sustainable Development, these guidelines encourage scientists to do three things: first, to engage with complementary knowledge systems and with citizen scientists, thereby capitalizing on all talents; second, to reflect critically on and take responsibility for the ethical implications of their research; and third, to call for policy adjustments and evidence-based solutions on the basis of their research.

It is a lot to ask. Many scientists just want to do their research and are wary of becoming science diplomats, contributing to a strong science-policy interface. The fact is, though, that we need scientists to help lay the foundation for a sustainable world. And as the challenges today and in the future are global and can be solved only at the global level, scientists need to keep pushing against political tendencies toward closing borders and thinking in nationalist answers only. Just like plants and nature itself, science knows no borders, and scientists need to collaborate with their colleagues in all corners of the world.

Johan Lange offered a fine example of borderless natural and scientific interchange in his introduction to a survey of Greenlandic flora in 1890. Echoing Asa Gray, Lange wrote, "Even though only a relatively low number of plants live in Greenland, the vegetation of this country is of no small plant-geographic interest. It forms a prominent part of the polar flora and offers instructive points of comparison between the northernmost part of mainland North America, and Iceland, Spidsbergen, Scandinavia and even the floras of countries as far away as northern Russia, Siberia, the Southern European Alps etc."[117]

As a botanist with extensive knowledge about plants growing in the geographical area of the world that has been in focus here, the last editor of *Flora Danica* had an interest in developing that "truly circumpolar cooperation" mentioned in the joint communique accompanying the official establishment in 1996 of the Arctic Council.

Conclusion

The 1970s are often remembered as a time of economic struggle and cultural change. In the Danish context, rising inflation and unemployment led to a number of labor conflicts, protesting the income policies of the government. Some of the most significant of these took place at the Royal Danish Porcelain Factory (today Royal Copenhagen) and involved the—mostly female—skilled workers painting Royal Copenhagen products, such as the blue fluted porcelain and the *Flora Danica* dinner set, entirely by hand. In 1973, the Danish government introduced equal pay and a forty-hour workweek, a major victory for the labor unions. For many female workers, though, equal pay remained a mere formality.

Inspired and supported by the women's movement, the popularity and power of which led to major cultural changes in Denmark in the 1970s, the female workers at Royal Copenhagen decided to take matters into their own hands. Affectionately known to the Danes as the "platter women," they went on strike twice to demand equal pay, once in 1972–1973 and again in 1976.[1] The 1976 strike, which lasted for thirteen weeks, between June and September, resulted in a great deal of sympathy as one highly visible part of it took place on the major walking street in Copenhagen just outside the Royal Copenhagen flagship store. Here, the platter women held public meetings to draw attention to their plight. When the weather permitted, they sat in the sunshine and painted paper plates with blue fluted and *Flora Danica* patterns for passersby to buy, earning some much-needed money during their strike. Each paper plate sold for 10 Danish kroner, today about seven dollars.

Much covered in the media, what is popularly known as "the strike of the platter women" in 1976 spurred a number of additional strikes and supportive actions. All around the country, solidarity meetings were held with women on strike making speeches while some of their colleagues painted paper plates for sale. In the end, they were successful. Though their employers at Royal Copenhagen did not agree to all their demands, the platter women did get a considerable pay raise. "Solidarity" was a key word for the striking women, whose labor fight gained the support of many members of the Danish population.

Painting on the spot for all to see and then selling those blue fluted and *Flora Danica* paper plates was a brilliant idea and a major PR stunt. The irony of transferring, as part of a labor conflict over equal pay, some of the most expensive porcelain patterns in the world on to paper plates, a cheap consumer good, was lost on no one. Today, a new *Flora Danica* porcelain plate sells for 12,800 Danish kroner (approximately $2,000), a teapot for 40,000 Danish kroner (approximately $6,300). But then, each piece has been handled by about thirty pairs of hands, painted three times, embellished twice with twenty-four-karat gold paint, and fired at least seven times.[2]

The Sociocultural Component of Class Struggles

The strike of the platter women was in many ways a classic European class struggle over wages. But there was more to it than money. The strikers were also engaged in a class struggle of a more social and cultural kind that concerned nondiscrimination and equal opportunities for women, broadly speaking. At a time of feminist mobilization, changing gender role attitudes, and women's issues becoming an important part of the political agenda, the striking women wanted to be taken seriously as equal democratic citizens and to be treated in a respectful and dignified manner.[3] They and their descendants, in Denmark as around the world, still do. To this day, gender inequality is apparent in educational and employment opportunity, or in health, for example, just as individual gender identity (e.g., male, female, or neither) is a major point of contention.

It is no wonder that gender equality is designated as a global priority of UNESCO. The organization "believes that all forms of discrimination based on gender are violations of human rights, as well as a significant barrier to the achievement of the 2030 Agenda for Sustainable Development and its 17 Sustainable Development Goals."[4] Gender is also a focus area for the mandate of the UN special rapporteur in the field of cultural rights and therefore figures prominently in many reports.[5] From a cultural rights perspective, the issue is one of furthering human creativity and learning, thereby equipping girls and women, along with boys and men, with the knowledge and skills necessary for enhancing their life choices: "Women and men must enjoy equal opportunities, choices, capabilities, power and knowledge as equal citizens."[6]

For the pursuit of the betterment of oneself as well as one's family and community, the right to science plays a prominent role. This is reflected in Farida Shaheed's reports. In her 2012 report on the right to science, she engages with the work of Indian-American anthropologist Arjun Appadurai.[7] Embracing what he calls "the politics of hope," Appadurai sees "the capacity to aspire as a social and collective capacity without which words such as 'empowerment,' 'voice,' and 'participation' cannot be meaningful."[8] If those who are disadvantaged, including women, do not aspire to a better life for themselves, their families, and their communities, they will not do anything to change and challenge the conditions of their own situation. But aspiring in itself is not enough to cause change. In order for such aspirations to be realized in practice, they have to be grounded in a wish to know more and to gain informed knowledge by means of what Appadurai calls the right to research:

> Without aspiration, there is no pressure to know more. And without systematic tools for gaining relevant new knowledge, aspiration degenerates into fantasy or despair. Thus, asserting the relevance of the right to research, as a human right, is not a metaphor. It is an argument for how we might revive an old idea, namely, that taking part in democratic society requires one to be informed. One can hardly be informed unless one has some ability to conduct research, however humble the question or however quotidian its inspiration. This is doubly true in a world where rapid change, new technologies and rapid flows of information change the playing field for ordinary citizens every day of the week.[9]

Picking up on Appadurai's link between the capacity to do research and the cultural capacity to "aspire," Shaheed discusses access to and participation in scientific research as an essential capacity for democratic citizenship. For women, as well as for everyone else, this capacity is a necessary precondition for real-life fights for better economic, political, and cultural conditions.

The underlying argument is that access to science (and culture) can be a major transformative tool for realizing and making dreams come true as well as for enjoying a life with dignity and meaningful participation in the community. Human dignity is a foundational value in international human rights law, serving variously as the grounds from which human rights are derived or as a constraint on the types of innovations and actions that are compatible with a human rights–based approach. Indeed, international human rights law expressly forbids any action or invention contrary to human dignity.[10]

Books, Reading, and Improving Life Chances

Appadurai's argument about the importance of gaining relevant new knowledge to make one's aspirations come true echoes the UNESCO Charter of the Book, issued on the occasion of the 1972 International Book Year.[11] Books and related material should, as the introduction to the charter states, "be accorded a position commensurate with the vital role they play in promoting individual fulfilment, social and economic progress, international understanding and peace."[12] In ten articles, the charter outlines the principles that should guide the treatment of books, both nationally and internationally. According to Article I, everyone has a "right to read" and to benefit from reading, and society must help people realize this right in practice. As "books are essential to education," (Article II), authors should be encouraged to be creative and initiatives should be taken to develop a sound publishing industry and infrastructure for book manufacturers, booksellers, libraries, and documentation services (Articles III to VIII). Finally, as "books constitute one of the major defences of peace because of their enormous influence in creating an intellectual climate of friendship and mutual understanding," restraints on international trade in books such as tariffs, taxes, and licenses should be reduced to a minimum (Articles IX and X).

But is it true that books and book-related resources enhance life chances? And if so, can this be measured with any accuracy? In a 2019 article, Joanna Sikoraa, M. D. R. Evans, and Jonathan Kelley answered both of these questions in the affirmative by documenting valuable effects of what they refer to as "scholarly culture" for adult literacy, adult numeracy, and adult technological problem solving.[13] They contrast arguments concerning "scholarly culture" as a social practice and way of life with the "cultural reproduction" and cultural capital arguments that build on the thinking of French sociologist Pierre Bourdieu. Bourdieu and his followers maintain that reading books is about status and cultural power. When well-off parents give their children books or take children to museums and art galleries, their intention is to give them tools with which they can master highbrow culture and signal class status to others. In priming "their offspring for socio-economic success regardless of their academic ability," the "cultural reproduction" argument goes, these elite parents "perpetuate educational inequalities."[14]

This is not what happens, though, Sikoraa and colleagues argue. Their findings, which are based on surveys conducted in thirty-one societies around the world between 2011 and 2015, show that access to cultural resources, chiefly books, in fact do not advantage the children of elite parents only, nor does access result in cultural reproduction that excludes everyone else. The beneficial effects of these cultural resources—vocabulary building, counterfactual thinking, and cognitive flexibility that enhances academic performance[15]—are strongest for people from the most disadvantaged homes: "bookishness as social practice, to which youth are acculturated, creates cognitive benefits which are not only immediate but also last into adulthood and are independent of educational and occupational standing (although bookishness also significantly enhances both forms of attainment)."[16]

The more books children and adolescents grow up with, the better. The larger the home library of the parents, the better their children do in school and in life in general. When they see their parents reading, they too want to read. Their cognitive skills, knowledge, and life-long tastes can also be stimulated by family social practices such as storytelling, imaginative play, or word games, just as they can learn from what they read in the books themselves.[17] Scholarly culture should be seen, Sikoraa and colleagues conclude, as a way of life—something children grow up with—and not as something that is cultivated to prime children for elite or highbrow life. Engaging with books of all kinds in childhood and adolescence "enhances adult competencies not only via attainments and regular reading habits in adulthood . . . but also directly, as a lifelong propensity to routinely include books into one's cultural and material environment."[18]

What will happen when children engage more with digital than with printed books? This will not automatically make books obsolete, Sikoraa, Evans, and Kelley speculate. The size of the home library children grow up with, their research shows, is related not only to higher levels of print literacy but also to digital literacy. For some time to come, reading books will therefore most likely continue to have a beneficial effect on adult competence with information and communications technologies.[19] In terms of expanding the book market and distributing books to many more children, digital books have considerable advantages over printed books. As digital devices become more powerful and less expensive, they will give access to reading materials to one billion children around the world who cannot get books in their native languages or cannot afford to buy books—"their best hope for escaping from poverty."[20]

Ending book hunger and putting books into the hands of all the world's children is crucial, argues Lea Shaver, in that "ensuring access to books is not only about improving educational outcomes and enhancing future job prospect. Stories also inspire, and may help an impoverished reader to envision optimistic narratives for his or her own life."[21] Shaver is one of the foremost scholars on cultural rights and the right to science, and I have quoted from her work several times in this book. Her work also served as an inspiration for Shaheed—especially on the issue of seeing the right to science as a cultural right and on emphasizing the importance of intellectual property for these rights.[22] In her important 2019 book, *Ending Book Hunger: Access to Print Across Barriers of Class and Culture*, Shaver highlights the need for nonprofit strategies and market, charitable, and government efforts to create incentives for the production of books. We should think about books as public benefits and goods just like education and health care, she contends.[23]

Because the stakes are so very high, a way must be found to encourage the publication of children's books in diverse languages at a lower cost and to make these available through digital libraries and digital devices such as mobile phones. Key to this endeavor, says Shaver, is the translation of books from one of the major languages, for example, English, into diverse languages. Not all stories are global enough for this, but some are: "Science and nature titles, animal stories, folk tales, and multicultural books."[24] Doing a survey of titles within each of these four categories that meet selection criteria for mass translation, such as to what extent book characters and settings reflect non-white, nonaffluent, or non-Western demographics, Shaver found at least seven thousand multicultural and science and nature titles, as well as stories about animal characters, that might work in many different languages. Together, she speculates, these may form a "new global canon of children's literature" that reflects the kinds of values that are well known in all cultures and that are promoted by the 2030 UN Sustainable Development Agenda: kindness, generosity, bravery, hard work, studiousness, and service to others.[25]

For the translation process to be successful, what Shaver calls "the copyright licensing dilemma" must be solved. Copyright law and licensing are a fundamental part of how commercial publishing works. So are translation rights.[26] But, suggests Shaver, there are ways of tailoring copyright law to the needs of nonprofit publishing. These include blanket licensing, fair use, copyright exceptions, and open licensing, and "once translation is accomplished, digital distribution scales up rapidly and inexpensively."[27] Making use of a mass translation and copyright strategy serving the needs of nonprofit publishing, Shaver concludes, "we can realistically create an adequate book supply for 95 percent of the world's children within a few years. Time is of the essence—and not just to reach the 2030 goal of all children learning well. . . . In the twentieth century, coordinated partnerships of charity, volunteers, government, and industry successfully eradicated smallpox and polio. Similar partnerships will be critical to the campaign to end book hunger. If we succeed, the effects will be transformative for one billion children now in school and for their communities."[28]

Children (and their parents) need books to learn and grow. Without reading materials they have less access to knowledge and cannot enjoy their right to benefit from, let alone participate in, scientific progress and its applications.

The right to science, as I have attempted to show in this book, is a little-known but potentially very powerful human right. One of four core cultural rights—the other three being the rights to education, to participate in cultural life, and authors' rights—it is outlined in UDHR Article 27 and repeated in ICESCR Article 15. Article 15(1) has three parts. It protects the right to participate in cultural life, the right to enjoy the benefits of scientific progress and its applications, or just the right to science, and authors' rights. Article 15(2) requires states to develop and disseminate science (and culture). As the drafters of the ICESCR knew, without dissemination and popularization there is no fulfillment of the right to science because otherwise people do not know about and cannot benefit from what is happening in the world of science. The third part of Article 15 concerns the obligation of states to respect the freedom of scientific (and cultural) research. Finally, Article 15(4) asks states to recognize the benefits of international contacts and cooperation in the scientific (and cultural) fields.

In all parts of Article 15, science and culture are mentioned side by side. Science is a part of culture and, like culture, it has to be shared. The free flow of people and ideas is crucial at a time where our most important problems—climate change, pandemics such as COVID-19, the growing gap between rich and poor—are global and have to be solved at the global level. A new Cold War–like atmosphere is currently engulfing science. The science that we need to solve world problems like pandemics and the challenges of climate change cannot be achieved without politically neutral agendas in which the global public good is paramount. One example is the current political tensions at the China–India border that have resulted in restrictions on scientific collaboration between those nations. "Science diplomacy on environmental issues in the Himalaya," write the authors of a 2020 letter in *Science*, "could increase the possibility of sustained peace along the international border and allow the two superpowers to lead the world toward a sustainable future. . . . We urge the governments of China and India to facilitate science diplomacy, starting with the Himalaya region."[29]

ICESCR Article 15 is rich with many interesting ways of getting out ideas that are highly topical right now. As I see it, this particular human right is about both basic and applied science. It is not just about, say, the right to health or the right to technology. It is about the right to hear and act on the kind of truth science can offer. Knowledge empowers.

NOTES

Introduction

1. Universal Declaration of Human Rights (1948), available at https://www.un.org/en/universal-declaration-human-rights/. International Covenant on Economic, Social and Cultural Rights (1966), available at https://www.ohchr.org/en/professionalinterest/pages/cescr.aspx.

2. I thank Dr. Peter Murray at the Max Planck Institute of Biochemistry in Martinsried, Germany, for sharing this example with me.

3. Audrey R. Chapman, "Towards an Understanding of the Right to Enjoy the Benefits of Scientific Progress and Its Applications," *Journal of Human Rights* 8, no. 1 (2009):1–36—here at p. 14.

4. Yvonne Donders, "Balancing Interests: Limitations to the Right to Enjoy the Benefits of Scientific Progress and Its Applications," *European Journal of Human Rights* 4 (2015): 486–503—here at p. 492.

5. *Report of the Special Rapporteur in the Field of Cultural Rights, Farida Shaheed. The Right to Enjoy the Benefits of Scientific Progress and Its Applications*, A/HRC/20/26, UN General Assembly, 2012, para. 43.

6. Ibid., para. 1.

7. Each of the monitoring bodies of the major human rights treaties publishes its interpretation of the provisions of its respective human rights treaty in the form of "general comments" or "general recommendations." Committee on Economic, Social and Cultural Rights, General Comment No. 25 on Science and Economic, Social and Cultural Rights, Articles 15(1)(b), 15(2), 15(3), and 15(4), E/C.12/GC/25, 7 April 2020.

8. Ibid., para. 10.

9. Ibid., paras. 8–11.

10. Ibid., para. 9. See also *Report of the Special Rapporteur in the Field of Cultural Rights, Farida Shaheed: The Right to Enjoy the Benefits of Scientific Progress and Its Applications*, A/HRC/20/26.

11. General Comment No. 25, para. 14.

12. Ibid., para. 13.

13. Ibid., paras. 77–84.

14. Helle Porsdam, *The Transforming Power of Cultural Rights: A Promising Law and Humanities Approach* (Cambridge, U.K.: Cambridge University Press, 2019).

15. Ibid.

16. See, e.g., Tara Smith, "Understanding the Nature and Scope of the Right to Science Through the *Travaux Préparatoires* of the Universal Declaration of Human Rights and the International Covenant on Economic, Social and Cultural Rights," *International Journal of Human Rights* 24, no. 8 (2020): 1156–1179.

17. See the mission and vision of the International Council of Science, created in 2018 as the result of a merger between the International Council for Science (ICSU) and the International Social Science Council (ISSC): https://council.science/about-us. Accessed on 15 August 2021.

18. Mary Ann Glendon, *Rights Talk: The Impoverishment of Political Discourse* (New York: Free Press, 1991). Helle Porsdam, *From Civil to Human Rights: Dialogues on Law and Humanities in the United States and Europe* (Cheltenham, U.K.: Edward Elgar, 2009).

19. Richard Pierre Claude, *Science in the Service of Human Rights* (Philadelphia: University of Pennsylvania Press, 2002).

20. Richard Pierre Claude, "Remarks to the AAAS Science and Human Rights Coalition Washington DC," July 23, 2009. Available at https://www.aaas.org/sites/default/files/RichardPierreClaude_Remarks.pdf. Accessed on 15 August 2021.

21. Ibid.

22. See, e.g., Anna Altman, "The Year of Hygge, the Danish Obsession with Getting Cozy," *New Yorker*, 18 December 2017. Available at https://www.newyorker.com/culture/culture-desk/the-year-of-hygge-the-danish-obsession-with-getting-cozy. Accessed on 15 August 2021.

23. I also did so in *The Transforming Power of Cultural Rights*.

Chapter 1

1. Philipp Blom, *Encyclopédie: The Triumph of Reason in an Unreasonable Age* (London: Fourth Estate, 2004), p. xvi. One of these kings, Assurbanipal (668–627 BCE), "kept a sizeable palace library (the remnants of which are now in the British Museum), and among the usual trading correspondence, inventories, and legends is a significant number of tablets containing lists of objects and names linked by theme, similarity, word root, or assonance."

2. Ibid., p. xvii.

3. Ibid., p. xviii.

4. Ibid., p. xvii.

5. These figures are Philipp Blom's (cover of his book *Encyclopédie*). *Encyclopeadia Britannica* talks about twenty-eight volumes, however. HTTPS://WWW.BRITANNICA.COM/TOPIC/ENCYCLOPEDIE. Accessed on August 15, 2021.

6. Quoted in Blom, *Encyclopédie*, p. 139.

7. Ibid., quoted p. 140. Italics in the original.

8. Ibid., cover.

9. Ibid., subtitle and p. 326.

10. Anna Lowenhaupt Tsing, *The Mushroom at the End of the World: On the Possibility of Life in Capitalist Ruins* (Princeton, N.J.: Princeton University Press, 2015), p. vii.

11. Henning Knudsen, *The Story Behind Flora Danica* (Copenhagen: Lindhardt og Ringhof in co-operation with Natural History Museum of Denmark, 2016), p. 9. The pages on the history of the *Flora Danica* in this and future chapters build on Knudsen's book as well as on the work of Peter Wagner, Ib Friis, and Jean Anker.

12. Ibid., quoted p. 10.

13. Georg Christian Oeder, quoted in Jean Anker, "From the Early History of the Flora Danica," *Libri* I (1951): 334–335—here at p. 345. My translation from the original German: "Wenn ich mein Urtheil am gelindesten ausdrücken soll, sehr unvollständig."

14. The full title of the work now colloquially and universally known as *Flora Danica* is *Icones plantarum sponte nascentium in regnis Daniae et Norvegiae, in ducatibus Slesvici et Holsatiae, et in comitatibus Oldenburgi et Delmenhorstiae, ad illustrandum opus de iisdem plantis,*

regio jussa exarandum, Florae Danicae nomine inscriptum. Hauniae. It sometimes alternatively appeared on the title page in Danish or German.

15. Anker, "From the Early History of the Flora Danica," p. 345.

16. Knudsen, *The Story Behind Flora Danica,* p. 105.

17. Ibid., p. 18.

18. Peter Wagner, *Til Publique nytte og brug. Bibliotheca Regia Historiae Naturalis 1752-1770* (Copenhagen: Lægeforeningens forlsag, 1994), p. 5.

19. See Jean Anker, "An Early Special Public Library," *Libri* 4, no. 1 (1953): 7–13—here at p. 12.

20. The royal Frederiks Hospital was Denmark's first hospital in the present-day meaning of the word. It was founded by King Frederik V and opened in 1757. It was situated in Bredgade in Copenhagen, in the building that currently houses Design Museum Denmark.

21. Wagner, *Til Publique nytte og brug,* p. 81.

22. Peter Wagner, "The Royal Work *Flora Danica,*" in *Flora Danica og det Danske hof. Udstilling af porcelæns-, guld- og sølvservice* (Copenhagen: Christiansborg Casle, 1990), pp. 15–44—here at p. 22.

23. Knudsen, *The Story Behind Flora Danica,* p. 21.

24. Wagner, "The Royal Work *Flora Danica,*" p. 22.

25. Quoted in Anker, "From the Early History of the Flora Danica," p. 344.

26. Wagner, "The Royal Work *Flora Danica,*" pp. 28–30.

27. Quoted in Anker, "From the Early History of the Flora Danica," p. 342.

28. Ibid., quoted p. 344. My rough translation from the German, which reads: "Meine grösste Absicht ist, die Krauter-Kenntnis gemeinnüziger zu machen, als sie würklich ist."

29. Quoted in Wagner, "The Royal Work *Flora Danica,*" p. 28.

30. Ibid.

31. Knudsen, *The Story Behind Flora Danica,* p. 21.

32. Wagner, "The Royal Work *Flora Danica,*" p. 42.

33. Knudsen, *The Story Behind Flora Danica,* p. 588.

34. Ibid., quoted p. 38.

35. Ibid., all quoted pp. 40–42.

36. Anker, "From the Early History of the Flora Danica," pp. 334–335.

37. Sital Kalantry, Jocelyn E. Getgen, and Steven Arrigg Koh, "Enhancing Enforcement of Economic, Social and Cultural Rights Using Indicators: A Focus on the Right to Education in the ECESCR," *Human Rights Quarterly* 32 (2010): 253–310—here at p. 263.

38. Universal Declaration of Human Rights, Article 27. Available at https://www.un.org/en/universal-declaration-human-rights/.

39. International Covenant on Economic, Social and Cultural Rights, Article 15,1(b). Available at https://www.ohchr.org/EN/professionalinterest/pages/cescr.aspx. Article 15(1)(c) concerns the right "to benefit from the protection of the moral and material interests resulting from any scientific, literary or artistic production of which he is the author."

40. UDHR Article 22 reads, "Everyone, as a member of society, has the right to social security and is entitled to realization, through national effort and international co-operation and in accordance with the organization and resources of each State, of the economic, social and cultural rights indispensable for his dignity and the free development of his personality."

41. Johannes Morsink, *The Universal Declaration of Human Rights: Origins, Drafting and Intent* (Philadelphia: University of Pennsylvania Press, 1999), p. 212.

42. Tara Smith, "Understanding the Nature and Scope of the Right to Science Through the Travaux Préparatoires of the Universal Declaration of Human Rights and the International Covenant on Economic, Social and Cultural Rights," *International Journal of Human Rights* 24, no. 8 (2020): 1156–1179.

43. Ibid.

44. *Report of the Special Rapporteur in the Field of Cultural Rights*, A/HRC/20/26, paragraph 21.

45. Some international lawyers would argue that the UDHR has by now become binding on all states through customary international law.

46. Annotations on the Text of the Draft International Covenants on Human Rights, UN Doc. A/2929, 1955—here at p. 7. Quoted in Henry J. Steiner and Philip Alston, *International Human Rights in Context; Law, Politics, Morals* (New York: Oxford University Press, 2000), p. 245.

47. See the wording of the Vienna Declaration and Programme of Action of 1993 that "All human rights are universal, indivisible and interdependent and interrelated." Available at https://www.ohchr.org/en/professionalinterest/pages/vienna.aspx.

48. I treat this issue in more detail in *The Transforming Power of Cultural Rights: A Promising Law and Humanities Approach* (Cambridge, U.K.: Cambridge University Press, 2019).

49. See, e.g., Katharine G. Young (ed.), *The Future of Economic and Social Rights* (Cambridge, U.K.: Cambridge University Press, 2019).

50. Charter of Fundamental Rights of the European Union (2000), Article 13. Available at https://www.europarl.europa.eu/charter/pdf/text_en.pdf.

51. General Comment No. 25 on Science and Economic, Social and Cultural Rights, Articles 15(1)(b), 15(2), 15(3), and 15(4), E/C.12/GC/25, 7 April 2020.

52. See, e.g., Klaus D. Beiter, "Where Have All the Scientific and Academic Freedoms Gone? And What Is 'Adequate for Science'? The Right to Enjoy the Benefits of Science and Its Applications," *Israel Law Review* 52, no. 2 (2019): 233–291.

53. Ibid. p. 238.

54. Naomi Oreskes, *Why Trust Science?* (Princeton, N.J.: Princeton University Press, 2019), pp. 68, 247.

55. I thank Dr. Peter Murray at the Max Planck Institute of Biochemistry in Martinsried, Germany, for suggesting this to me. Other scholars are working on ethical guidelines for the computer, AI, and neuroengineering industries—see, e.g., calls for a "technocratic oath" by the NeuroProtection group in Chile (https://nri.ntc.columbia.edu/projects, accessed on August 15, 2021) and the call for a so-called tech pledge, originating in Denmark (https://www.techpledge .org, accessed on August 15, 2021).

56. Audrey R. Chapman, "Towards an Understanding of the Right to Enjoy the Benefits of Scientific Progress and Its Applications," *Journal of Human Rights* 8, no. 1 (2009): 1–36. Available from https://doi.org/10.1080/14754830802701200.

57. See, e.g., M. Chandler, L. See, K. Copas, A. M. Bonde, B. C. López, F. Danielsen, et al., "Contribution of Citizen Science Towards International Biodiversity Monitoring," *Biological Conservation* 213 (2017): 280–294.

58. General Comment No. 25, para. 11.

59. Ibid.

60. A/HRC/20/26, para. 7.

61. Ibid., para. 43.

62. Ibid., para. 52. See also General Comment No. 25 on Science, Section III, concerning elements of the right and limitations.

63. AAAS Statement on Scientific Freedom and Responsibility, *Science* 358, no. 6362 (October 27, 2017): 462.

64. *Intersections of Science, Ethics and Human Rights: The Question of Human Subjects Protection.* Report of the Science Ethics and Human Rights Working Group, AAAS Science and Human Rights Coalition, 2012, p. 2. Available at https://www.aaas.org/sites/default/files/s3fs-public/reports/ScienceEthicsHumanRights.pdf. Accessed on August 15, 2021.

65. Kalantry et al., "Enhancing Enforcement of Economic, Social and Cultural Rights Using Indicators," p. 255.

66. AAAS Science and Human Rights Coalition: About. Available at https://www.aaas.org/page/coalition-about. Accessed on August 15, 2021.

67. See reports and resources by the AAAS Science and Human Rights Coalition. Available athttps://www.aaas.org/coalition-resources. Accessed on August 15, 2021.

68. Margaret Weigers Vitullo and Jessica Wyndham, *Defining the Right to Enjoy the Benefits of Scientific Progress and Its Applications: American Scientists' Perspectives.* AAAS Science and Human Rights Coalition, 2013.

69. J. M. Wyndham, M. W. Vitullo, K. Kraska, N. Sianko, P. Carbajales, C. Nuñez-Eddy, and E. Platts. *Giving Meaning to the Right to Science: A Global and Multidisciplinary Approach.* Report prepared under the auspices of the AAAS Scientific Responsibility, Human Rights and Law Program and the AAAS Science and Human Rights Coalition, 2017.

70. *Venice Statement on the Right to Enjoy the Benefits of Scientific Progress and Its Applications.* UNESCO, 2009. Available at http://unesdoc.unesco.org/images/0018/001855/185558e.pdf. Accessed on August 15, 2021.

71. Jessica M. Wyndham, Margaret W. Vitullo, Rebecca Everly, Teresa M. Stoepler, and Nathaniel Weisenberg, "The Right to Science – From Principle to Practice and the Role of National Science Academies," in Helle Porsdam and Sebastian Porsdam Mann (eds.), *The Right to Science: Then and Now* (New York: Cambridge University Press, 2021), pp. 211–230.

72. L. Segal et al., "Promoting Human Rights Through Science. *Science* 358, no. 6359 (2017): 34–47.

73. Morsink, *The Universal Declaration of Human Rights*, pp. 210–212. See also Porsdam, *The Transforming Power of Cultural Rights*, Chapter 3, pp. 67–95.

74. Farida Shaheed, "The United Nations Cultural Rights Mandate: Reflections on the Significance and Challenges," in Lucky Belder and Helle Porsdam (eds.), *Negotiating Cultural Rights: Issues at Stake, Challenges and Recommendations* (Cheltenham, U.K.: Edward Elgar), pp. 21–36—here at p. 22.

75. See the bibliography of *The Transforming Power of Cultural Rights for* an update on the latest scholarship on cultural rights, including the right to science.

76. See,e.g.,StevenPinker,"WhyWeAreNotLivinginaPost-TruthEra:An(Unnecessary)Defense of Reason and a (Necessary) Defense of Universities' Role in Advancing it." *Skeptic* 24, no. 3 (2019). Available at https://www.skeptic.com/reading_room/steven-pinker-on-why-we-are-not-living-in-a-post-truth-era/?gclid=EAIaIQobChMIkO2q7Pr25QIVB7LtCh2E5AmYEAAYASAAEgKI3PD_BwE. Accessed on August 15, 2021.

77. Oreskes, *Why Trust Science?*, p. 245.

78. Ibid., p. 129.

79. Ibid., p. 128.

80. Ibid., pp. 141, 149, 127.

81. Ibid., p. 152.

82. Ibid., p. 129.

83. See Daniel Oberhaus, "Why Some Citizens Reject Science: Bridging the Gulf to Science deniers," *Harvard Magazine*, September/October 2021. Available at https://www .harvardmagazine.com/2021/09/right-now-clash-science-ideology?utm_source=email&utm _medium=newsletter&utm_term=monthly&utm_content=SO21&utm_campaign=081821. Accessed on 19 August 2021.

84. Ibid., p. 154.

85. Ibid., p. 153.

86. Ibid., pp. 152–153.

87. Ibid., p. 138.

88. 2017 UNESCO Recommendation on Science and Scientific Researchers, 1(d)(i), 1(a) (ii). Available at http://portal.unesco.org/en/ev.php-URL_ID=49455&URL_DO=DO_TOPIC& URL_SECTION=201.html.

89. Ibid., 1(a)(i).

90. General Comment No. 25, paras. 4–7.

91. "Voltaires Epistel til Kongen af Danmark i Anledning af Trykkefriheden, der blev tilladt i alle hans Stater 1771" (Voltaire's epistle to the King of Denmark on the occasion of freedom of the press being allowed in all His states in 1771). My translation from the Danish, which was translated from the French by the Danish poet and critic Knud Lyne Rahbek and was published in the journal *Minerva* in October 1797, pp. 75–83. Available in Danish at https://da.wikisource .org/wiki/Voltaires_Epistel_til_Kongen_af_Danmark. The emphases are Rahbek's own.

92. "Christian VII," Kings' Collection, Rosenborg, http://www.kongernessamling.dk/en /rosenborg/person/christian-vii.

93. Per Olov Enquist, *The Royal Physician's Visit* (New York: Overlook Press, 2001).

94. Wagner, "The Royal Work *Flora Danica*," pp. 30, 32.

95. Ibid. The color plates and illustrations are known as the "Icones florae Danicae."

96. See, e.g., Haochen Sun, *Reinvigorating the Human Right to Technology, Michigan Journal of International Law* 41, no. 2 (2020): 279. Available at https://repository.law.umich.edu/mjil /vol41/iss2/3.

97. I participated as an invited speaker in "Panel 2: Relationship with Other Rights." A general summary of the event may be found at https://www.ohchr.org/EN/NewsEvents/Pages /DisplayNews.aspx?NewsID=23714&LangID=E.

98. General Comment No. 25, para. 8.

99. Aurora Plomer, *Patents, Human Rights and Access to Science* (Cheltenham, U.K.: Edward Elgar, 2015), p. 165.

Chapter 2

1. Tara Smith argues, for example, that "the right to science has become so indelibly attached to the right to culture in both the UDHR and the ICESCR is unfortunate. . . . Permanently subsuming the right to science under the banner of cultural rights, both ideologically and textually in these two key human rights instruments, may have blunted the effect and the perception of the right over time." See Tara Smith, "Understanding the Nature and Scope of the Right to Science Through the *Travaux Préparatoires* of the Universal Declaration of Human Rights and

the International Covenant on Economic, Social and Cultural Rights," *International Journal of Human Rights* 24 (2020): 1156–1179.

2. One very interesting example is Ib Friis, "G. C. Oeder's Conflict with Linnaeus and the Implementation of Taxonomic and Nomenclatural Ideas in the Monumental Flora Danica Project (1761–1883)," *Gardens' Bulletin Singapore* 71 (Suppl. 2) (2019): 53–85. Another is Henning Knudsen, *Fortællingen om Flora Danica* (Copenhagen: Lindhardt og Ringhof, 2014). The English version, *The Story Behind Flora Danica*, was published by the same publisher in 2016 and is one of the sources I use in this book.

3. One example is Atlas Flora Danica, a crowdsourcing project carried out between 1992 and 2012, which resulted in a publication in three volumes as well as a website. Per Hartvig, *Atlas Flora Danica*, three volumes (Copenhagen: Danish Botanical Society and Gyldendal, 2016). This was the second time that the Danish Botanical Society took the initiative to map the wild Danish flora. The first time this happened was in 1904—an effort that was not completed until 1980. In mapping the wild Danish flora, the monitoring work was done on a voluntary basis by members of the Danish Botanical Society. The purposes bear a striking resemblance to those of the original work: "For this project, Denmark was divided into 2228 squares of 5 × 5 km. In all investigated squares, all plant species were recorded, with exact location. About every other square were surveyed and thereby obtaining good coverage of the whole country. . . . Atlas Flora Danica's purpose is to: 1) Show the Danish plant species' current distribution and regional frequency, including decline or spread of species. Causes of these changes are discussed. 2) Examine the current status of red-listed, protected and rare plant species. 3) Publicize the Danish flora and the importance of protecting the plants and their habitats." Botanical Society of Denmark, Atlas Flora Danica, https://www.gbif.org/dataset/8df9af24-1fbd-4699-a545-9a2867ff25fe #description. Accessed on August 15, 2021.

4. Peter Wagner, "Icones Florae Danicae. Flora Danicas 'Urteteignere' og 'Illuminationsskolen for Quindekiønnet," in Vibeke Woldbye (ed.), *Blomster fra Sans og Samling* (Copenhagen: Rhodos, 1990), pp. 93–100—here at p. 100. See "Conclusions" in Chapter 1 for more on this topic.

5. There was furthermore a time when Royal Copenhagen sold Flora Danica perfume, https://www.parfumo.net/Perfumes/Royal_Copenhagen/Flora_Danica. Accessed on August 15, 2021.

6. See, e.g., Helle Porsdam, *The Transforming Power of Cultural Rights: A Promising Law and Humanities Approach* (New York: Cambridge University Press, 2018).

7. American Declaration of the Rights and Duties of Man Article XIII, first paragraph. See Cesare P. R. Romano, "The Origins of the Right to Science: The American Declaration on the Rights and Duties of Man," in Helle Porsdam and Sebastian Porsdam Mann (eds.), *The Right to Science: Then and Now* (New York: Cambridge University Press, 2021), pp. 33–53.

8. Richard Pierre Claude, *Science in the Service of Human Rights* (Philadelphia: University of Pennsylvania Press, 2002), p. 15.

9. "At its core, Article 15 requires that science be used as an instrument for human benefit, and that the process of doing scientific research and the development of applications from that science be consistent with fundamental human rights principles such as nondiscrimination and equal treatment, participation and transparency in decision-making, and free and informed consent to participation in research." AAAS, "Right to Science: FAQs." Available at https://www .aaas.org/programs/scientific-responsibility-human-rights-law/resources/faqs. Accessed on August 15, 2021.

10. S. P. Mann, Y. Donders, C. Mitchell, V. J. Bradley, M. F. Chou, M. Mann, G. Church, and H. Porsdam. Opinion: Advocating for Science Progress as a Human Right. *Proceedings of the National Academy of Science of the United States of America* 115, no. 43 (2018): 10820–10823.

11. George Edwards, quoted in Judith Magee, *Art of Nature: Three Centuries of Natural History Art from Around the World* (London: Natural History Museum, 2009), p. 6.

12. Ibid.

13. The Danish West India Guinea Company had annexed Saint. Thomas in 1672 and Saint John in 1718, and Saint Croix was purchased from the French West India Company in 1733. The Danish Company went bankrupt in 1754, and the Danish king assumed control of the islands in 1755. See Poul E. Olsen (ed.), *Danmark og kolonierne: Vestindien—St. Croix, St. Thomas og St. Jan* (Copenhagen: Gad, 2017).

14. Knudsen, *The Story Behind Flora Danica*, p. 105.

15. Daniela Bleichmar, "Painting as Exploration: Visualizing Nature in Eighteenth Century Colonial Science," *Colonial Latin American Review* 15, no. 1 (June 2006): 81–104—here at p. 84.

16. Magee, *Art of Nature*, p. 11.

17. Ibid.

18. Ibid., p. 201.

19. Ibid.

20. Ibid., p. 199. In *Sex, Botany and Empire: The Story of Carl Linnaeus and Joseph Banks* (London: Icon Books, 2017), Patricia Fara writes how Banks today "is little known in his native country but has become Australia's founding father. . . . Banks, botany and Australia are indissolubly tied together, even though he was only there for a few weeks—and even then, often at sea rather than on land" (pp. 155–56). But in *The Multifarious Mr. Banks: From Botany Bay to Kew, the Natural Historian Who Shaped the World* (New Haven, Conn.: Yale University Press, 2020), Toby Musgrave tells of Banks's profound impact on Britain as the longest-serving president of the Royal Society and advisor to King George III.

21. Peter Wagner, *Til publique nytte og brug: Bibliotheca Regia Historiae Naturalis, 1752-1770* (Copenhagen: Lægeforeningens forlag, 1994–1995), pp. 87–88.

22. Peter Wagner, "Haver, blomster og billedbøger: Strøtanker om mode set fra et botanikersynspunkt," in Vibeke Woldbye (ed.), *Blomster fra Sans og Samling* (Copenhagen: Rhodos, 1990), pp. 9–24—here at p. 19.

23. Magee, *Art of Nature*, p. 202.

24. Wagner, "Icones Florae Danicae," p. 96.

25. Friis, "G. C. Oeder's Conflict with Linnaeus and the Implementation of Taxonomic and Nomenclatural Ideas in the Monumental Flora Danica Project (1761–1883)," p. 56.

26. Ibid, quoted in Appendix 1, "Oeder's Invitation to Subscribe to a New Work Called *Flora Danica* (1761)," p. 78.

27. Wagner, "Icones Florae Danicae," pp. 94–95.

28. Ibid., p. 95. My own translation from the Danish.

29. Ibid., p. 96.

30. Ibid., p. 98.

31. Ibid., pp. 99–100.

32. See, e.g., "Flora Danica: Science in Bloom on Delicate Porcelain," https://kongeligeslotte.dk/en/palaces-and-gardens/christiansborg-palace/explore-christiansborg-palace/flora-danica.html. Accessed on August 15, 2021.

33. Steen Nottelmann, "Flora Danica and the Royal Copenhagen Porcelain Manufactury," in Ole Villumsen Krog (ed.), *Flora Danica og det danske hof* (Copenhagen: Christiansborg Castle, 1990), pp. 175–194—here at p. 182.

34. Ibid., p. 186.

35. Ibid., p. 182.

36. Ibid., p. 186.

37. See "The Dinner Set," Royal Library, http://www5.kb.dk/en/tema/floradanica/stellet .html. Accessed on August 15, 2021.

38. See "The Stories of Flora Danica," Royal Copenhagen, http://floradanica.royalcopen hagen.com/the-craft/. Accessed on August 15, 2021.

39. Ibid.

40. The Danish Asger Jorn Museum has a copy of the drawing in which cabbage forms the hair of a person, https://www.museumjorn.dk/da/pierre_alechinsky__flora_danica.asp. Accessed on August 15, 2021.

41. See https://snm.ku.dk/SNMnyheder/alle_nyheder/2013/2013.2/danske_kunstnere_giver _flora_danica_nyt_liv/ (in Danish). Accessed on August 15, 2021.

42. See, e.g., https://snm.ku.dk/snmnyheder/alle_nyheder/2013/2013.5/flora_danica_zoom/ (in Danish). Accessed on August 15, 2021.

43. Knudsen, *Fortællingen om FLORA DANICA*. English version, *The Story behind FLORA DANICA*.

44. Royal Copenhagen, Flora, https://www.royalcopenhagen.com/us/en_US/Series/Flora/c /Flora. Accessed on August 15, 2021.

45. Flora D, the story of Flora Danica, https://flora-danica.com/. The Flora D designers have found "inspiration in the great Danish work Flora Danica, which is one of our nation's greatest contributions to the Enlightenment." Accessed on August 15, 2021.

46. https://floradanica.eu/pages/about-us. Accessed on August 15, 2021.

47. Flora Danica Marguerite, http://www.brille-mie.dk/smykker/floradanica/floradanica .asp. Georg Jensen Daisy, https://www.georgjensen.com/europe/jewellery/daisy. Accessed on August 15, 2021.

48. New Nordic Food, Nordic Council of Ministers, https://www.norden.org/en/information /new-nordic-food-manifesto. Accessed on August 15, 2021.

49. Sam George, "Cultivating the Botanical Woman: Rousseau, Wakefield and the Instruction of Ladies in Botany," *Zeitschrift für Pädagogische Historiographie* 12, no. 1 (2006): 3–11— here at p. 3.

50. Ibid.

51. Ibid., p. 10.

52. Fara, *Sex, Botany & Empire*, p. 44: "Botany was sexy, dangerous—and big business." See also p. 17: "Historians often neglect the 18th century because it lacks famous figureheads such as Isaac Newton or Charles Darwin, yet this was a crucial period when science started to become established and gain prestige."

53. George, "Cultivating the Botanical Woman," p. 10.

54. UNESCO Institute for Statistics, https://en.unesco.org/news/just-30-world%E2%80 %99s-researchers-are-women-whats-situation-your-country. Accessed on August 15, 2021.

55. "With science 'held back by a gender gap,' Guterres calls for more empowerment for women and girls," *UN News*, February 10, 2020. Available at https://news.un.org/en/story/2020 /02/1057111. Accessed on 15 August 2021.

56. UNESCO, Priority Gender Equality, https://en.unesco.org/genderequality. Accessed on 15 August 2021.

57. As a specialized agency of the UN, UNESCO participated in the negotiation and the adoption of Agenda 2030. Especially involved in developing SGD 4 (quality education), UNESCO also explicitly declared itself willing to work toward harnessing science, technology, innovation, and knowledge for sustainable development ("UNESCO moving forward the 2030 Agenda for Sustainable Development." Available at http://unesdoc.unesco.org/images/0024 /002477/247785e.pdf). Accessed on August 15, 2021.

With its promise of "leaving no one behind," the 2030 Agenda notably places human rights at the center of all UN-related action, and such an approach automatically underscores the need to abolish inequalities of opportunities on account of discrimination. With respect to science, this means, as a 2017 UNESCO publication puts it, that it is important "to remediate past inequalities and patterns of exclusion, actively encourage women and persons of other under-represented groups to consider careers in sciences, and endeavor to eliminate biases against women and persons of other under-represented groups in work environments and appraisal" [UNESCO Recommendation on Science and Scientific Researchers, 2017, Article 13(c)].

58. See Information on the Mandate, (e), http://www.ohchr.org/EN/Issues/CulturalRights /Pages/MandateInfo.aspx. Accessed on August 15, 2021. This paragraph is taken from Chapter 7 in Porsdam, *The Transforming Power of Cultural Rights*, p. 189.

59. Report of the special rapporteur in the field of cultural rights, A/67/287, 10 August 2012.

60. Naomi Oreskes, *Why Trust Science?* (Princeton, N.J.: Princeton University Press, 2019), pp. 49–51. This quotation can be found on p. 137.

61. Information on the mandate, (e).

62. In her report celebrating the tenth anniversary of the mandate, Bennoune writes that while "the mandate has been very successful in addressing gender, with two dedicated reports, as well as the mainstreaming of gender and the cultural rights of lesbian, gay, bisexual, transgender and intersex persons in both thematic and country work . . . much more needs to be done on the cultural rights of persons with disabilities." Report of the special rapporteur in the field of cultural rights, *Cultural Rights: Tenth Anniversary Report*, A/HRC/40/53, 2019, paras. 86, 87.

63. CESCR General Comment No. 25 (2020) on Science and Economic, Social and Cultural Rights, Articles 15(1)(b), 15(2), 15(3), and 15(4), E/C.12/GC/25, paras. 34, 35. The other groups in need of special protection mentioned here are persons living in poverty and indigenous peoples.

64. 2017 UNESCO Recommendation on Science and Scientific Researchers, III.

65. Ibid.

66. Ibid., 16(b)(iii).

67. Charter of Fundamental Rights of the European Union, 2012/C 326/02, Article 17. Available at https://eur-lex.europa.eu/legal-content/EN/TXT/?uri=CELEX:12012P/TXT. Accessed on August 15, 2021.

68. For the United States, see https://www.copyright.gov/title17/. Accessed on August 15, 2021.

69. See Porsdam Mann et al., "Opinion: Advocating for Science Progress as a Human Right."

70. There are other important restrictions, for example, university, institute, and government restrictions placed on sharing knowledge and research. Some universities encourage patenting anything and everything.

71. CESCR General Comment No. 25 on Science, paras. 58–62.

72. See Christine Mitchell, "Epilogue: Tensions in the Right to Science Then and Now," in Porsdam and Porsdam Mann (eds.), *The Right to Science: Then and Now*, pp. 286–297: "Most science is done in academic and government settings or for-profit companies. Scientists who share their findings in scholarly articles, monographs, and books for the academic market are generally not paid to publish, and their data, findings and publications are usually not freely available to the general public. Instead, scientists submit their work to journals and publishing corporations, such as Elsevier, who handle scientific peer review and publication, and the information is privatized and put behind a paywall. Libraries then pay an institutional fee for faculty to have access, or individuals may pay a fee per article. Sometimes authors pay publishing companies a fee as well, usually in the thousands of dollars, for their work to be freely available—typically labelled 'open access'—in which case the for-profit publishing company collects fees from both the universities and authors." Here at p. 294.

73. CESCR General Comment No. 25 on Science, paras. 58–62.

74. This is noted in several papers in the literature survey done for our PNAS article; see Porsdam Mann et al., "Opinion: Advocating for Science Progress as a Human Right."

75. Aurora Plomer, "IP Rights and Human Rights: What History Tells Us and Why It Matters," in Helle Porsdam and Sebastian Porsdam Mann (eds.), *The Right to Science: Then and Now*– here at p. 56.

76. Ibid.

77. See Chapter 6 on "Copyright, Patents, Authors' Rights, and the Right to Culture and Science" in my *The Transforming Power of Cultural Rights* for a more in-depth discussion of these issues.

78. Reports of the special rapporteur in the field of cultural rights, A/HRC/20/26 (2012), A/HRC/28/57 (2014), and A/70/279 (2015). These reports are available on the website of the special rapporteur in the field of cultural rights, under Documents.

79. Report of the special rapporteur in the field of cultural rights, A/70/279, Summary.

80. Ibid., para. 32.

81. A/HRC/28/57, para. 48.

82. Ibid., para. 49.

83. CESCR General Comment No. 25 on Science, para. 62.

84. Plomer, "IP Rights and Human Rights," p. 58.

85. Smith (2020), "Understanding the Nature and Scope," p. 8.

86. See, e.g., my interview with Farida Shaheed, "UCPH Talking About—Negotiating Cultural Rights," 17 November 2015, https://video.ku.dk/ucph-talking-about-negotiating. Accessed August 15, 2021.

87. Farida Shaheed, "The United Nations Cultural Rights Mandate: Reflections on the Significance and Challenges," in Lucky Belder and Helle Porsdam (eds.), *Negotiating Cultural Rights: Issues at Stake, Challenges and Recommendations* (Cheltenham, UK: Edward Elgar, 2017), p. 26.

88. Personal conversations with concerned scientists.

89. Report of the special rapporteur in the field of cultural rights, *Universality, Cultural Diversity and Cultural Rights*, A/73/227, 25 July 2018, paras. 47, 25.

90. Ibid., para. 15.

91. Report of the special rapporteur in the field of cultural rights, A/HRC/34/56, 16 January 2017, paras. 2, 22.

92. Malala Yousafzai was awarded the Nobel Peace Prize in 2014 for her struggle for the right of all children, especially girls, to education. In her Nobel lecture, she spoke of education as

"one of the blessings of life." See Yousafzai, Malala. "Nobel Lecture 2014," http://www.nobelprize .org/nobel_prizes/peace/laureates/2014/yousafzai-lecture_en.html. Accessed on August 15, 2021. Yousafzai was speaking from experience. She was shot by the Taliban on her way home from school in October 2012 and today lives in England. I tell her story as part of Chapter 3 of my *The Transforming Power of Cultural Rights*.

93. Report of the special rapporteur in the field of cultural rights, A/HRC/34/56, Summary.

94. Julie B. Zimmerman, Paul T. Anastas, Hanno C. Erythropel, and Walter Leitner, "Designing for a Green Chemistry Future," *Science* 367, no. 6476 (2020): 397–400.

95. Colin Poitras, "Green Chemistry Is Vital to a Sustainable Future, YSPH Professor Says," Yale School of Public Health, Office of Public Relations, 10 February 2020. Available at https:// ysph.yale.edu/news-article/green-chemistry-is-vital-to-a-sustainable-future-ysph-professor -says/. Accessed on August 15, 2021.

96. See, e.g., Klaus D. Beiter, "Where Have All the Scientific and Academic Freedoms Gone? And What Is 'Adequate for Science'? The Right to Enjoy the Benefits of Science and Its Applications," *Israel Law Review* 52, no. 2 (2019): 233–291.

97. There were also other reasons. Prominent British scientists had lobbied for a while to have science be a part of UNESCO alongside education and culture. See chapter 5 in Porsdam, *The Transforming Power of Cultural Rights*, for a more in-depth discussion.

98. Ellen Wilkinson, quoted in Roberto Andorno, "The role of UNESCO in promoting universal human rights: From 1948 to 2005," in C. Caporale and I. Pavone (eds.), *International Biolaw and Shared Ethical Principles. The Universal Declaration on Bioethics and Human Rights* (Oxford: Routledge), 2018, pp. 7–21—here at p. 8.

99. European Commission, Funding and Tender Opportunities, https://ec.europa.eu /research/participants/docs/h2020-funding-guide/cross-cutting-issues/ethics_en.htm. Accessed August 15, 2021.

100. See Valerie J. Bradley, "Implications of the Right to Science for People with Disabilities," in Porsdam and Porsdam Mann (eds.), *The Right to Science*, pp. 150–165. This paragraph builds on Bradley's article.

101. Ibid.

102. C. P. Snow, "The Two Cultures. The Rede Lecture, 1959," in C. P. Snow, *The Two Cultures* (Cambridge, U.K.: Cambridge University Press, 1998). For a longer and more thorough discussion of this, see Helle Porsdam, "Digital Humanities: On Finding the Proper Balance Between Qualitative and Quantitative Ways of Doing Research in the Humanities," *Digital Humanities Quarterly* 7, no. 3 (2013). Available at http://www.digitalhumanities.org/dhq/vol/7 /3/000167/000167.html.

103. Marita Akjhøj Nielsen (ed.), *Det fremmede som historisk drivkraft Danmark efter 1742. Et festskrift til Hendes Majestæt Dronning Margrethe II ved 70-års-fødselsdagen den 16. april 2010* (Copenhagen: Det Kongelige Danske Videnskabernes Selskab). My translation from the Danish. Available in Danish only at https://botaniskforening.dk/wp-content/uploads/2015/12/botanik _dronningesymposium.pdf.

104. Jacob Abbott, *Margaret of Anjou* (Frankfurt am Main: Outlook Verlag, 2019), p. 54. This is a reproduction of the original from 1861.

105. Marita Akhøj Nielsen, *Det fremmede som historisk drivkraft Danmark efter 1742*, p. 9.

106. Daisy Meaning and Symbolism, https://www.ftd.com/blog/share/daisy-meaning-and -symbolism. Accessed on August 15, 2021.

107. Richard Mabey, *Weeds: The Story of Outlaw Plants* (London: Profile Books, 2012), pp. 109–110.

108. Hans Christian Andersen, "The Butterfly" (1861), https://andersen.sdu.dk/vaerk /hersholt/TheButterfly_e.html. Accessed on August 15, 2021.

109. See https://vimeo.com/94368001, which also features short videos with the participating artists.

110. See Farida Shaheed, "The United Nations Cultural Rights Mandate," in Belder and Porsdam (eds.), *Negotiating Cultural Rights*, pp. 21–36.

Chapter 3

1. See the mission and vision of the International Council of Science, created in 2018 as the result of a merger between the International Council for Science and the International Social Science Council, https://council.science/about-us. Accessed on August 16, 2021.

2. Article 15(2) International Covenant on Economic, Social and Cultural Rights,. Available at https://www.ohchr.org/en/professionalinterest/pages/cescr.aspx.

3. Peter Wagner, *Til Publique nytte og brug. Bibliotheca Regia Historiae Naturalis 1752–1770* (Copenhagen: Lægeforeningens forlsag, 1994), p. 5.

4. See B. Graf and V. Rodekamp (eds.), *Museen zwischen Qualität und Relevanz. Denkschrift zur Lage der Museen* (Berlin: G + H Verlag, 2012).

5. Kim Sloan, "'Aimed at Universality and Belonging to the Nation': The Enlightenment and the British Museum," in Kim Sloan and Andrew Burnett (eds.), *Enlightenment: Discovering the World in the Eighteenth Century* (London: British Museum Press, 2003), p. 13.

6. Quoted in Sloan, p. 13.

7. Wagner, *Til Publique nytte og brug: Bibliotheca Regia Historiae Naturalis 1752–1770*, p. 9.

8. Ibid., p. 10.

9. Quoted in Peter Wagner, "En disputatshandling og dens følger," *Bibliotek for læger*, Årg. 184, h. 2 (1992), p. 149. My translation from the Danish.

10. Ibid., p. 159.

11. Ibid., p. 164.

12. Quoted in Wagner, *Til Publique nytte og brug*, p. 15. My translation from the Danish.

13. Ibid., p. 22.

14. Ibid., p. 21.

15. Jean Anker, "An Early Special Public Library," *Libri* 4, no. 1 (1953): 7–13— here at p. 12.

16. Peter Wagner, "The Royal Work *Flora Danica*," in *Flora Danica og det Danske hof. Udstilling af porcelæns-, guld- og sølvservice* (Copenhagen: Christiansborg Casle, 1990), pp. 15–44—here at p. 22.

17. Wagner, *Til Publique nytte og brug*, p. 111.

18. Ibid., p. 42.

19. Ibid., p. 116.

20. Naomi Oreskes, *Why Trust Science* (Princeton, N.J.: Princeton University Press, 2019), p. 151.

21. Peter Wagner, "Georg Christian Oeder's medicinske iagttagelser paa hans rejser i Norge" (Copenhagen: Bibliotek for læger, 1994), p. 150.

22. Ibid., pp. 150–151.

23. Ibid., pp. 153–155.

24. Thomas Hager, "How One Daring Woman Introduced the Idea of Smallpox Inoculation to England," *Time*, March 5, 2019. Available at https://time.com/5542895/mary-montagu-smallpox/. Accessed on August 16, 2021.

25. "Smallpox," The History of Vaccines: An Educational Resource by the College of Physicians of Philadelphia. – Available at https://www.historyofvaccines.org/content/articles/history-smallpox. Accessed on August 16, 2021.

26. World Health Organization, "Ten Threats to Global Health in 2019." Available at https://www.who.int/news-room/spotlight/ten-threats-to-global-health-in-2019. Accessed on August 16, 2021.

27. Wagner, "Georg Christian Oeder's medicinske iagttagelser paa hans rejser i Norge," p. 158.

28. Ibid.

29. Ibid., p. 161. My translation from the Danish.

30. Ibid., p. 163.

31. Ibid.

32. "Vision and values," Natural History Museum of Denmark. Available at https://snm.ku.dk/english/vision-and-values/. Accessed on August 17, 2021.

33. "Citizen science," Natural History Museum of Denmark. Available at https://www.nhm.ac.uk/take-part/citizen-science.html. Accessed on August 17, 2021.

34. Quoted in Bjarne Nielsen, "Mit bedste ledelsesråd." Available at http://nielsencompany.dk/ArtiklerTilDownloadHjemmeside/050406ARHJPKlummeBedsteLedelsesraad.pdf Accessed on August 17, 2021. My translation from Danish.

35. Cécile Petitgand, Catherine Régis, and Jean-Louis Denis, *Is Science a Human Right? Implementing the Principle of Participatory, Equitable, and Universally Accessible Science* (Ottawa: Canadian Commission for UNESCO's Idealab, 2019), pp. 6–7.

36. Hala Sharkas, "Translation Quality Assessment of Popular Science Articles Corpus Study of the Scientific American and Its Arabic Version," *trans-kom* 2, no. 1 (2009): 42–62—here at p. 42. Available at http://www.trans-kom.eu/bd02nr01/trans-kom_02_01_03_Sharkas_Translation_Quality_Assessment.20090721.pdf. Accessed on August 17, 2021.

37. Report of the UN special rapporteur in the field of cultural rights, A/HRC/20/26, para. 37.

38. Petitgand et al., "Is Science a Human Right?," p. 6.

39. Ibid., e.g.

40. CESCR General Comment No. 25 (2020) on Science and Economic, Social and Cultural Rights, Articles 15(1)(b), 15(2), 15(3), and 15(4), E/C.12/GC/25. See also Farida Shaheed's 2012 report on the right to enjoy the benefits of scientific progress and its application A/HRC/20/26, subsection 4, paras. 45–48.

41. CESCR General Comment No. 25, para. 15.

42. Ibid., para. 16.

43. Ibid., para. 17. In paragraph 28, the General Comment lists women; persons with disabilities; lesbian, gay, bisexual, transgender, and intersex persons; indigenous peoples; and persons living in poverty as "groups that have experienced systemic discrimination in the enjoyment of the right to participate in and to enjoy the benefits of scientific progress and its applications." For reasons of space, however, the comment focuses on women, persons with disabilities, persons living in poverty, and indigenous peoples.

44. Ibid., paras. 17–18.

45. Ibid. para. 19.

46. Ibid., para. 20. Paragraph 13 lays out in more detail how the CESCR sees freedom of scientific research.

47. Ibid., para. 46.

48. As Lea Bishop Shaver puts it, "To call science a human right is precisely to insist that the supply of scientific knowledge and the development of technology must not be left entirely – perhaps not even primarily—to market forces." Shaver, Lea Bishop, *The Right to Science: Ensuring that Everyone Benefits from Scientific and Technological Progress*, Indiana University Robert H. McKinney School of Law Research Paper No. 2015-5 (February 12, 2015), p. 7. Available at https://ssrn.com/abstract=2564222.

49. General Comment No. 25 on science, paras. 53–55.

50. L. Pettibone, et al., *Citizen Science for All: A Guide for Citizen Science Practitioners*, Bürger Schaffen Wissen (GEWISS). Available at https://www.buergerschaffenwissen.de/sites/default/files/assets/dokumente/handreichunga5_engl_web.pdf. Accessed on August 17, 2021.

51. Eva Stratilová Urválková and Svatava Janoušková, "Citizen Science—Bridging the Gap Between Scientists and Amateurs," *Chemistry Teacher International* 1, no. 2 (2019).

52. Shaver, "The Right to Science," p. 5.

53. Petitgand et al., "Is Science a Human Right?," p. 6.

54. Grundtvig Programme, European Commission. Available at https://web.archive.org/web/20130726040701/http://ec.europa.eu/education/lifelong-learning-programme/grundtvig_en.htm. Accessed on August 17, 2021.

55. *The Lifelong Learning Programme: Education and Training Opportunities for All*. European Commission. Available at https://web.archive.org/web/20130726030319/http://ec.europa.eu/education/lifelong-learning-programme/doc78_en.htm. Accessed on August 17, 2021.

56. More information about the Danish folk high schools may be found at https://ipc.dk/about/what-is-a-danish-folk-high-school. Accessed on February 22, 2020.

57. An interesting longevity study on life expectancy in the United States, published in 2020, seems to establish a link between education, health, and longevity. For the first time in decades, life expectancy is in decline in the United States among middle-aged adults. Combining data about 5,114 black and white individuals in four U.S. cities, recruited for the study when they were in their early twenties and now in their mid-fifties, the Coronary Artery Risk Development in Young Adults (CARDIA) study suggests that improving access to quality education can help change this troubling decline. Public health scholars have speculated that inaccessible health care and other socioeconomic factors along with race might be behind the problem, but as it turns out, lower education level is an independent predictor of longevity, even after accounting for the effects of other variables such as income and race. Strikingly, "when looking at race and education at the same time, differences related to race all but disappeared: 13.5 percent of black subjects and 13.2 percent of white subjects with a high school degree or less died during the course of the study. By contrast, 5.9 percent of black subjects and 4.3 percent of whites with college degrees had died." Quoted in Bill Hathaway, "Want to Live Longer? Stay in School, Study Suggests," *Yale News*, February 20, 2020. Available at https://news.yale.edu/2020/02/20/want-live-longer-stay-school-study-suggests?utm_source=YaleToday&utm_medium=Email&utm_campaign=YT_YaleNews%20-%20Alumni%20From%20Peoplehub_2-21-2020. Accessed on August 17, 2021. The study referred to is Brita Roy, Catarina I. Kiefe, David R. Jacobs, David C. Goff, Donald Lloyd-Jones, James M. Shikany, Jared P. Reis, Penny Gordon-Larsen, and Cora E. Lewis, "Education, Race/Ethnicity, and Causes of Premature Mortality Among Middle-Aged Adults in 4 US Urban Communities: Results from CARDIA, 1985-2017," *American Journal of Public Health* 110, no. 4 (April 2020): 530–536.

58. Johannes Morsink, *The Universal Declaration of Human Rights: Origins, Drafting & Intent* (Philadelphia: University of Pennsylvania Press, 1999), p. 112.

59. Quoted in Tara Smith (2020), "Understanding the Nature and Scope of the Right to Science Through the *Travaux Préparatoires* of the Universal Declaration of Human Rights and the International Covenant on Economic, Social and Cultural Rights," *International Journal of Human Rights* 24, no. 8 (2020): 1156–1179—here, p. 22, note 92.

60. Quoted in William A. Schabas, "Looking Back: How the Founders Considered Science and Progress in their Relation to Human Rights," *European Journal of Human Rights* 2015, no. 4 (2015): 504–518—here at p. 505.

61. Ibid.

62. Quoted in ibid., pp. 506–507. See also Aurora Plomer, "IP Rights and Human Rights: What History Tells Us and Why It Matters," in Helle Porsdam and Sebastian Porsdam Mann (eds.). *The Right to Science: Then and Now* (New York: Cambridge University Press, 2021), pp. 54–75.

63. See Chapter 5 in my *The Transforming Power of Cultural Rights* for a more in-depth treatment of this issue.

64. Jaime Clifton-Ross, Ann Dale, and Robert Newell, "Frameworks and Models for Disseminating Curated Research Outcomes to the Public," *SAGE Open* (April-June 2019): 1–13.

65. B. Graf and V. Rodekamp (eds.), *Museen zwischen Qualität und Relevanz. Denkschrift zur Lage der Museen* (Berlin: G + H Verlag, 2012).

66. See, e.g., I. Lawley, "Local Authority Museums and the Modernizing Government Agenda in England," *Museum and Society* 1, no. 2 (2003): 75–86; R. Sandell, "Social Inclusion, the Museum and the Dynamics of Sectoral Change," *Museum and Society* 1, no. 1 (2001): 45–62.

67. B. J. Pine and J. H. Gilmore, *The Experience Economy: Work Is Theater & Every Business a Stage* (Boston: Harvard Business School Press, 1999).

68. Scott G. Paris, "Situated Motivation and Informal Learning," *Journal of Museum Education* 22, no. 2–3 (1997): 22–27.

69. S. Brown, "A Critique of Generic Learning Outcomes," *Journal of Learning Design* 2, no. 2 (2007): 22–30.

70. H. Jenkins, *Fans, Bloggers, and Gamers: Exploring Participatory Culture* (New York: New York University Press, 2006).

71. Chern Li Liew, "Participatory Cultural Heritage: A Tale of Two Institutions' Use of Social Media," *D-Lib Magazine* 20, no. 3/4 (2014). Available at http://www.dlib.org/dlib/march14/liew/03liew.html. Accessed on August 17, 2021.

72. See, e.g., L. E. Whiteley, A. Stenslund, K. Arnold, and T. Söderqvist. "'The House' as a Framing Device for Public Engagement in STEM Museums," *Museum & Society* 15, no. 2 (2017): 217–235.

73. Ibid., pp. 218–219.

74. Clifton-Ross et al., "Frameworks and Models for Disseminating Curated Research Outcomes to the Public."

75. Ibid. p. 2.

76. Ibid.

77. Ibid., p. 7.

78. Amber Dance, "How Museum Work Can Combine Research and Public Engagement," *Nature* 12 December 2017. Available at https://www.nature.com/articles/d41586-017-08458-0. Accessed on August 17, 2021.

79. Ibid.

80. Karl Ove Knausgård, "In the Black Forest with the Greatest Living Artist," *New York Times Magazine*, February 12, 2020. A version of this article appeared in print on February 16, 2020, page 26, with the title "In Search of Anselm Kiefer."

81. Medical Museion. The Culture of Medicine—Yesterday, Today, Tomorrow. www .museion.ku.dk.

82. Mind the Gut. https://www.museion.ku.dk/mindthegut/. Accessed on August 17, 2021.

83. Writing from the Gut—Writing Workshop. https://www.museion.ku.dk/2019/06/writing -from-the-gut-writing-workshop/. Accessed on August 17, 2021.

84. Event: Meditate with Microbes. https://www.museion.ku.dk/2019/05/event-mediter -med-mikrober/ (in Danish). Accessed on August 17, 2021.

85. Ken Arnold, Adam Bencard, Bente Vinge Pedersen, Thomas Söderqvist, and Karin Tybjerg, "A House of Collaboration: Investigating the Intersections of Art and Biomedicine," in Camilla Rossi-Linnemann and Giolia de Martini (eds.), *Art in Science Museums: Towards a Post-Disciplinary Approach* (London: Routledge, 2020), pp. 48–60—here at p. 57.

86. Ibid.

87. Florian Heigl, Barbara Kieslinger, Katharina T. Paul, Julia Uhlik, and Daniel Dörler, "Opinion: Toward an International Definition of Citizen Science," *Proceedings of the National Academies of Science of the United States of America* 116, no. 17 (2019): 8089–8092.

88. Ibid.

89. M. Andrejevic, "Social Network Exploitation," in Z. Papachrissi (ed.), *A Networked Self: Identity, Community, and Culture on Social Network Sites* (New York: Routledge, 2011), pp. 82–102.

90. Rick Bonney, Jennifer L. Shirk, Tina B. Phillips, Andrea Wiggins, Heidi L. Ballard, Abraham J. Miller-Rushing, and Julia K. Parrish, "Next Steps for Citizen Science," *Science* 343, no. 6178 (2014): 1436–1437.

91. Petitgand et al., "Is Science a Human Right?," p. 15.

92. Bonney et al., "Next Steps for Citizen Science."

93. Petitgand et al., "Is Science a Human Right?," p. 15.

94. Morsink, *The Universal Declaration of Human Rights*, p. 215.

95. Peter Aronsson and Gabriella Elgenius, "Introduction: Making Museums and Nations," in Peter Aronsson and Gabriella Elgenius (eds.), *National Museums and Nation-Building in Europe, 1750-2010: Mobilization and Legitimacy, Continuity and Change* (London: Routledge, 2015), pp. 1–9—here at p. 1.

96. Neal MacGregor, "To Shape the Citizens of 'That Great City, the World,'" in James Cuno (ed.), *Whose Culture? The Promise of Museums and the Debate over Antiquities* (Princeton, N.J.: Princeton University Press, 2009), p. 40.

97. See, e.g., Anthony Shelton, "Critical Museology: A Manifesto," *Museum Worlds: Advances in Research* 1 (2013): 7–23. Available at https://arenet.org/img/Critical%20Museology %20A%20Manifesto.pdf#page=2&zoom=auto,792,944. Accessed on August 17, 2021.

98. Museum for the United Nations–UN Live, https://museumfortheun.org/about/. Accessed on August 17, 2021.

99. Ibid.

100. My Mark: My City, https://mymarkmycity.org/. Accessed on August 8, 2021.

101. Judy Diamond, "Inclusion and Relevance in Natural History Museums," in Hooley McLaughlin and Judy Diamond (eds.), *Science Museums in Transition: Unheard Voices* (London: Routledge, 2020), p. 3.

102. Marita Akjhøj Nielsen (ed.), *Det fremmede som historisk drivkraft Danmark efter 1742. Et festskrift til Hendes Majestæt Dronning Margrethe II ved 70-års-fødselsdagen den 16. april 2010* (Copenhagen: Det Kongelige Danske Videnskabernes Selskab). Available at https://botaniskforening.dk/wp-content/uploads/2015/12/botanik_dronningesymposium.pdf.

103. Ib Friis, "Hvor dansk er den danske flora?," ibid., pp. 13–25—here at p. 14.

104. Quoted in Helge Kragh, "Mellem provinsialisme og internationalisme. Dansk naturvidenskab i oplysningstiden," ibid., pp. 26–34—here at p. 31. Freely translated from the Danish.

105. Ibid.

Chapter 4

1. See James Fallows, "The 50 Greatest Breakthroughs Since the Wheel," *The Atlantic*, November 2013. Available at https://www.theatlantic.com/magazine/archive/2013/11/innovations-list/309536/. Accessed on August 17, 2021.

2. International Covenant on Economic, Social and Cultural Rights, Article 15(3). Available at https://www.ohchr.org/en/professionalinterest/pages/cescr.aspx.

3. General Comment No. 17: The Right of Everyone to Benefit from the Protection of the Moral and Material Interests Resulting from any Scientific, Literary or Artistic Production of Which He or She is the Author (Article 15, Paragraph 1(c) of the ICESCR), E/C.12/GC/17, para. 4.

4. "EU budget 2021-2027: Commission Proposal to Further Strengthen Climate Action," https://ec.europa.eu/clima/policies/budget/mainstreaming_en. Accessed on August 17, 2021.

5. Another example is ethnozoology, which is today very interesting in light of the practices in Chinese wet markets.

6. Henning Knudsen, *The Story Behind Flora Danica* (Copenhagen: Lindhardt and Ringhof, 2016), p. 47.

7. Ibid., p. 252.

8. New Nordic Cuisine (Copenhagen: Nordic Council of Ministers, 2008), p. 15. Available at http://norden.diva-portal.org/smash/get/diva2:701317/fulltext01.pdf. Accessed on August 17, 2021.

9. Ib Friis, "G.C. Oeder's Conflict with Linnaeus and the Implementation of Taxonomic and Nomenclatural Ideas in the Monumental Flora Danica Project (1761–1883)," *Gardens' Bulletin Singapore* 71, Suppl. 2 (2019): 53–85.

10. Paul L. Farber, "Buffon and the Concept of Species, " *Journal of the History of Biology* 5, no. 2 (1972): 259–284. "Nature progresses by unknown gradations and consequently does not submit to our absolute divisions when passing by imperceptible nuances, from one species to another and often from one genus to another," Buffon wrote in 1749, for example (Buffon, quoted in Friis, "G.C. Oeder's Conflict with Linnaeus," p. 63). A few years later, Buffon made his preference for a biological species concept more explicit: "Each species—each succession of individuals that can successfully reproduce with each other—will be considered as a unit and treated separately. We will not use families, genera, orders, and classes, any more than nature does. A species, then, is only a constant succession of similar individuals that can reproduce together" (ibid.).

11. Ibid.

12. Mathias Persson, "Building an Empire in the Republic of Letters: Albrecht von Haller, Carolus Linnaeus, and the Struggle for Botanical Sovereignty," *Circumscribere* 17 (2016): 18–40—here at pp. 23–24.

13. See Friis, "G. C. Oeder's Conflict with Linnaeus," which presents new translations of these letters.

14. Ibid., p. 67.

15. Ibid., p. 53. It is not entirely clear why later *Flora Danica* editors Müller and Vahl decided to follow Oeder's and not Linnaeus's principles. Vahl grew up in Bergen and was a student of Linnaeus before he came to Copenhagen. Like Müller and Vahl, Hornemann was obviously aware of the debate between Oeder and Linnaeus. In 1823, he started presenting reports at the meetings of the Royal Danish Academy of Sciences and Letters on the scientifically interesting issues of new *Flora Danica* fascicles, a tradition that subsequent editors kept up. In one of these reports, Hornemann pragmatically wrote that he had introduced changes "in nomenclature due to the demand of the times and the changing views of the various editors" (quoted in Friis, p. 70).

16. "Imposing order on a lawless proliferation of names, Linnaeus created a system 250 years ago that, with substantial modification, still stands today," writes Jay Withgott, for example. "Is It 'So Long, Linnaeus'? In High-Stakes Name Game, Phylogenetic Nomenclature Puts Its Chips on the Table," *BioScience* 50, no. 8 (August 2000): 646–651.

17. Friis, "G. C. Oeder's Conflict with Linnaeus," p. 72.

18. Withgott, "Is It 'So Long, Linnaeus'?"

19. Ibid.

20. I thank Ib Friis for making this observation (personal communication).

21. Kathy J. Willis, "Introduction to the State of the World's Plants 2017," in Kathy J. Willis (ed.), *State of the World's Plants 2017* (London: Royal Botanic Gardens, Kew), p. 2. Available at https://stateoftheworldsplants.org/2017/report/SOTWP_2017.pdf. Accessed on March 18, 2020.

22. Ibid.

23. Bob Allkin, "Useful Plants – Medicines: At Least 28,187 Plant Species are Currently Recorded as Being of Medicinal Use," ibid., pp. 22–28. Available at https://stateoftheworldsplants .org/2017/report/SOTWP_2017_4_useful_plants_medicines.pdf. Accessed on March 18, 2020.

24. Ibid.

25. Ibid.

26. See, e.g., the Marie Curie Initial Training Network (ITN), MedPlant, which between 2013 and 2017 trained young researchers "in new phylogenetic approaches for exploration of medicinal plant diversity as well as complimentary and entrepreneurial skills relevant for work in the pharmaceutical industry, regulatory bodies, NGOs and academia." MedPlant produced guides to local medicinal plants that were handed out to the communities involved, as wells as a practical manual on conducting and communicating ethnobotanical research: https://cordis .europa.eu/project/id/606895/reporting. Accessed on August 17, 2021.

27. Brøndegaard, quoted in Henning Knudsen, "Florafortællinger til nytte og fornøjelse – Vagn J. Brøndegaard." In Håkan Tunón (ed.), *Etnobotanik. Planter i skik og brug, i historien og folkemedicinen. Vagn J. Brøndegaards biografi, bibliografi og artikler i udvalg på dansk* (Centrum för biologisk mångfald, Uppsala & Kungl. Skogs- och Lantbruksakademien, Stockholm), pp. 21–26—here at p. 22. My translation from the Danish.

28. Vagn J. Brøndegaard, "Vegetabilske kontraceptiva," reprinted in Håkan Tunón (ed.), *Etnobotanik. Planter i skik og brug*, pp. 630, 629. My translation from the Danish.

29. Vagn Brøndegaard, *Flora og fauna: dansk etnobotanik*, Vol. I–IV (Copenhagen: Rosenkilde and Bagger, 1978–1980). Brøndegaard also wrote on zoology and other issues. Here, I am merely concerned with his botanical work. Lisa Sennerby-Forsse, "Med Brøndegaard, KSLA ock LSU i Linnés fotspår – Ett akademiföretal," ibid., pp. 9–10.

30. Vagn J. Brøndegaard, "Danish children's games with flowers and other parts of plants," reprinted in Håkan Tunón (ed.), *Etnobotanik. Planter i skik og brug*, pp. 701–780.

31. John C. Ryan, "Cultural Botany: Toward a Model of Transdisciplinary, Embodied, and Poetic Research into Plants," *Nature and Culture* 6, no. 2 (2011): 123–148.

32. Ibid., pp. 143, 129, 130.

33. See, e.g., R. Monastersky, Anthropocene: The Human Age. *Nature* 519 (2015): 144–147.. E. O. Wilson wrote in 2017, for example, that, "Even as our species destroys the natural world at an accelerating rate, nature remains a source of deep love and fear. As we hasten to blanket Earth with a humanized environment, we should—we must—pause to consider how and why our relation to nature exists. That degree of self-understanding can be achieved only by a blending of science and the humanities." E. O. Wilson, *The Origins of Creativity* (New York: Liveright, 2017), p. 127.

34. Committee on Economic, Social and Cultural Rights, Discussion Paper, Day of General Discussion, 9 October 2018, question 16. Available at https://www.ohchr.org/Documents /HRBodies/CESCR/Discussions/2018/discussionpaper.pdf. Accessed on August 17, 2021.

35. General Comment No. 17: The Right of Everyone to Benefit from the Protection of the Moral and Material Interests Resulting from any Scientific, Literary or Artistic Production of Which He or She Is the Author (Article 15, Paragraph 1(c) of the ICESCR), E/C.12/GC/17, para. 4.

36. "The Committee focuses primarily on the 'right of everyone to enjoy the benefits of scientific progress and its applications' (art. 15.1.b), as it is the right most frequently invoked in relation to science. However, the purpose of this General Comment is not confined to this right, but also to develop the relationship more broadly between science and ESCRs. The Committee will also examine the other elements of article 15 related to science, especially the obligations of State parties to take steps for the conservation, development and diffusion of science (art 15.2.), to respect the freedom indispensable for scientific research (15.3) and to promote international contacts and co-operation in the scientific field (15.4)." General Comment No. 25, para. 3.

37. General Comment No. 25, para. 45.

38. Ibid., para. 40.

39. Ibid., para. 41.

40. Ibid., para. 4.

41. 2017 UNESCO Recommendation on Science and Scientific Researchers, 39 C/73, Annex III, Paragraph 1(a)(ii).

42. General Comment No. 25, para. 6.

43. Report of the special rapporteur in the field of cultural rights, *The Right to Enjoy the Benefits of Scientific Progress and Its Applications*. A/HRC/20/26, May 14, 2012, para. 24.

44. A/HRC/20/26, para. 7.

45. Farida Shaheed, "The United Nations Cultural Rights Mandate: Reflections on the Significance and Challenges," in Lucky Belder and Helle Porsdam (eds.), *Negotiating Cultural Rights: Issues at Stake, Challenges and Recommendations* (Cheltenham, U.K.: Edward Elgar, 2017), pp. 21–58—here at p. 32.

46. A/HRC/20/26, para. 23.

47. Ibid., para. 21.

48. This and the following two paragraphs are based on W. P. Metzger, "Academic Freedom and Scientific Freedom," *Daedalus* 107, no. 2 (1978): 93–114, and on Sebastian Porsdam Mann,

Maximilian M. Schmid, Peter V. Treit, and Helle Porsdam, "Scientific Freedom: A Constitutive Element of the Right to Science," in preparation.

49. Article 5, Basic Law for the Federal Republic of Germany in the revised version published in the Federal Law Gazette Part III, classification number 100-1, as last amended by Article 1 of the Act of 28 March 2019 (Federal Law Gazette I p. 404). Available in English at https://www.gesetze-im-internet.de/englisch_gg/englisch_gg.html#p0034.

50. Debbie Sayers, "Article 13—Freedom of the Arts and Sciences," in Steve Peers, Tamara Hervey, Jeff Kenner, and Angela Ward (eds.), *The EU Charter of Fundamental Rights: A Commentary* (London: Hart/Beck, 2014), pp. 379–400.

51. *Handyside v. UK* (1976), quoted ibid., p. 382. Unlike the United States, all EU member states have ratified or acceded to the ICESCR.

52. Klaus D. Beiter, "Where Have All the Scientific and Academic Freedoms Gone? And What Is 'Adequate for Science'? The Right to Enjoy the Benefits of Science and Its Applications," *Israel Law Review* 52, no. 2 (2019): 233–291—here at pp. 238, 243.

53. Ibid., p. 244.

54. Sayers, "Article 13—Freedom of the Arts and Sciences," p. 385.

55. General Comment No. 13 (1999) on the right to education, E/C.12/1999/10, paragraph 38. Available at https://www.refworld.org/pdfid/4538838c22.pdf. Accessed on 25 March 2020.

56. Beiter, "Where Have All the Scientific and Academic Freedoms Gone?," p. 247.

57. UNESCO Recommendation Concerning the Status of Higher-Education Teaching Personnel, 11 November 1997. Available at http://portal.unesco.org/en/ev.php-URL_ID=13144& URL_DO=DO_TOPIC&URL_SECTION=201.html. Accessed on 25 March 2020. In this recommendation, academic freedom is defined as "the right, without constriction by prescribed doctrine, to freedom of teaching and discussion, freedom in carrying out research and disseminating and publishing the results thereof, freedom to express freely their opinion about the institution or system in which they work, freedom from institutional censorship and freedom to participate in professional or representative academic bodies" (para. 26).

58. Report of the special rapporteur on the promotion and protection of the right to freedom of opinion and expression, A/75/261 (2020). In this report, the special rapporteur writes that, "Although there are many ways in which the freedom of opinion and expression protects and promotes academic freedom, there is no single, exclusive international human rights framework for the subject" (para. 5). Academic freedom, the rapporteur continues, "should be understood to include the freedom of individuals, as members of academic communities (e.g., faculty, students, staff, scholars, administrators and community participants) or in their own pursuits, to conduct activities involving the discovery and transmission of information and ideas, and to do so with the full protection of human rights law" (para. 8).

59. General Comment No. 25, para. 14.

60. Declaration on the Human Genome and Human Rights (11 November 1997), http:// portal.unesco.org/en/ev.php-URL_ID=13177&URL_DO=DO_TOPIC&URL_SECTION=201 .html; Universal Declaration on Bioethics and Human Rights (19 October 2005), https://en .unesco.org/themes/ethics-science-and-technology/bioethics-and-human-rights. Both accessed on August 17, 2021. See also Sayers, "Article 13—Freedom of the Arts and Sciences," p. 385.

61. General Comment No. 25, para. 1.

62. Yvonne Donders, "Balancing Interests: Limitations to the Right to Enjoy the Benefits of Scientific Progress and Its Applications," *European Journal of Human Rights* 4 (2015): 486–503— here at p. 492.

63. General Comment No. 25, para. 41.

64. This is an important theme in the 2012 special rapporteur report on the right to science, for example.

65. General Comment No. 25, para. 29.

66. Information on the mandate of the UN special rapporteur in the field of cultural rights: https://www.ohchr.org/EN/Issues/CulturalRights/Pages/MandateInfo.aspx. Accessed on August 17, 2021. General Comment No. 25 also mentions LGBTI among the groups for whom temporary special measures might be necessary. However, "owing to limitations of space, this General Comment focuses on women, persons with disabilities, persons in poverty and indigenous peoples," General Comment No. 25, para. 29.

67. General Comment No. 25, para. 37.

68. Ibid., paras. 51–52. See also Yvonne Donders, "The Right to Enjoy the Benefits of Scientific Progress: In Search of State Obligations in relation to Health," *Medicine, Health Care and Philosophy* 14, no. 4 (2011): 371–381.

69. Article 2(1) of the ICCPR provides that each state party will undertake to respect and ensure the rights recognized in ICCPR, whereas Article 2 of the ICESCR provides that the state party will take steps *to use its maximum available resources gradually to realize the rights* of the covenant. The two articles of the two covenants thus impose different types of obligations, those under ICESCR being more conduct oriented and contextual. This has in the past made many argue that the ICESCR can be taken less seriously than the ICCPR (see Chapter 1).

70. CESCR General Comment No. 3: The Nature of States Parties' Obligations (Article 2, Paragraph 1 of the covenant), E/1991/23. Available at http://www.refworld.org/docid /4538838e10.html. This and the following two paragraphs are based on Sebastian Porsdam Mann, Helle Porsdam and Yvonne Donders, "'Sleeping Beauty': The Right to Science as a Global Ethical Discourse," *Human Rights Quarterly* 42 (2020): 332–356—here at pp. 341–342.

71. Ibid., para. 1.

72. Ibid., para. 2.

73. General Comment No. 25, para. 25.

74. See, e.g. Klaus D. Beiter, "Where Have All the Scientific and Academic Freedoms Gone?"

75. Christian Starck, "Freedom of Scientific Research and Its Restrictions in German Constitutional Law," *Israel Law Review* 39, no. 2 (2006): 110–126.

76. Catherine Rhodes and John Sulston, "Scientific Responsibility and Development," *European Journal of Development Research* 22, no. 1 (February 2010): 3–9. Available at https://www .ncbi.nlm.nih.gov/pmc/articles/PMC5580797/. Accessed on August 17, 2021.

77. Paul Berg, "Asilomar 1975: DNA Modification Secured," *Nature* 455 (2008): 290–291.

78. An important part of this self-governance was a voluntary agreement by scientists not to conduct certain kinds of experiments considered too high-risk, such as the cloning of highly pathogenic organisms or organisms carrying toxin genes. Scientists also agreed to carry out less risky experiments under certain conditions meant to reduce the biological risks involved.

79. Berg, "Asilomar 1975."

80. Chronologically, the legal and judicial development of this principle first began in the U.S. with the Occupational Safety and Health Act of 1970. See Marco Bocchi, "Is the EU Really More Precautionary than the US? Some Thoughts in Relation to TTIP Negotiations," *EJIL: Talk!,* August 9, 2016, https://www.ejiltalk.org/is-the-eu-really-more-precautionary-than-the -us-some-thoughts-in-relation-to-ttip-negotiations/. Accessed on August 17, 2021.

81. Didier Bourguignon, *The Precautionary Principle: Definitions, Applications and Governance*, European Parliamentary Research Service, European Union, 2016, p. 1. Available at https://www.europarl.europa.eu/RegData/etudes/IDAN/2015/573876/EPRS_IDA(2015)573876_EN.pdf. Accessed on August 17, 2021.

82. The one proposed by UNESCO, which is repeated in the General Comment on science, covers "harm to humans or the environment that is threatening to human life or health, or serious and effectively irreversible, or inequitable to present or future generations, or imposed without adequate consideration of the human rights of those affected. The judgement of plausibility should be grounded in scientific analysis. Analysis should be ongoing so that chosen actions are subject to review." In *The Precautionary Principle*, UNESCO World Commission on the Ethics of Scientific Knowledge and Technology, , 2005, p. 14. Available at https://unesdoc.unesco.org/ark:/48223/pf0000139578. Accessed on November 5, 2019.

83. Bourguignon, *The Precautionary Principle* (2016), p. 1.

84. Vagn J. Brøndegaard, "Sagnet om Sambucus ebulus" (The Legend of Sambucus ebulus), reprinted in Håkan Tunón (ed.), *Etnobotanik. Planter i skik og brug, i historien og folkemedicinen. Vagn J. Brøndegaards biografi, bibliografi og artikler i udvalg på dansk.* Centrum för biologisk mångfald, Uppsala & Kungl. Skogs- och Lantbruksakademien, Stockholm, 2015, pp. 833–835—here at 833. My translation from the Danish.

85. Knudsen, *The story behind Flora Danica*, p. 244.

86. Hans Christian Andersen, "The Elder-Tree Mother." Available at https://andersen.sdu.dk/vaerk/hersholt/TheElderTreeMother_e.html. Accessed on August 17, 2021.

87. Brøndegaard, "Sagnet om Sambucus ebulus."

88. Ibid., p. 834. My translation from the Danish.

89. Ibid., p. 833. My translation from the Danish.

90. Retraction Watch: Tracking Retractions as a Window into the Scientific Process, https://retractionwatch.com. The website publishes a freely available table and database. Accessed on August 17, 2021.

91. Roberto Andorno, "The Right to Science and the Evolution of Scientific Integrity," in Helle Porsdam and Sebastian Porsdam Mann (eds.), *The Right to Science: Then and Now* (New York: Cambridge University Press, 2021), pp. 91–103.

92. Responsible Research and Innovation. Horizon 2020, European Commission, https://ec.europa.eu/programmes/horizon2020/en/h2020-section/responsible-research-innovation. Accessed on August 17, 2021.

93. General Comment on Science, para. 61.

94. Ibid., para. 22.

Chapter 5

1. "WHO Director-General's opening remarks at the media briefing on COVID-19, 11 March 2020, https://www.who.int/dg/speeches/detail/who-director-general-s-opening-remarks-at-the-media-briefing-on-covid-19---11-march-2020. Accessed on 1August 18, 2020.

2. "COVID-19 Pandemic: Humanity Needs Leadership and Solidarity to Defeat the Corona Virus." United Nations Development Programme, https://www.undp.org/coronavirus. Accessed on August 18, 2021.

3. Statement on the coronavirus disease (COVID-19) pandemic and economic, social and cultural rights. E/C.12/2020/1, April 6, 2020, para. 19.

4. Ibid., para. 23.

5. International Covenant on Economic, Social and Cultural Rights, Article 15(4). Available at https://www.ohchr.org/en/professionalinterest/pages/cescr.aspx.

6. Balachandar Vellingiria, Kaavya Jayaramayyab, Mahalaxmi Iyerb, Arul Narayanasamy-c,Vivekanandhan Govindasamyd, Bupesh Giridharane, Singaravelu Ganesang, Anila Venugopala, Dhivya Venkatesana, Harsha Ganesana, Kamarajan Rajagopalana, Pattanathu K. S. M. Rahmanh, Ssang-Goo Choi, Nachimuthu Senthil Kumarj, and Mohana Devi Subramaniam, "COVID-19: A Promising Cure for the Global Panic," *Science of the Total Environment*725 (10 July 2020): 138277, https://doi.org/10.1016/J.SCITOTENV.2020.138277. Accessed on August 18, 2021.

7. Jun-ling Ren, Ai-Hua Zhang, and Xi-Jun Wang, "Chinese Medicine for COVID-19 Treatment," *Pharmacological Research* 155 (2020): 104743, https://doi.org/10.1016/j.phrs.2020.104743. Accessed on August 18, 2021.

8. CESCR General Comment No. 25, paragraph 6. See also 2017 UNESCO Recommendation on Science and Scientific Researchers, 39 C/73, Annex III, paragraph 1(a)(ii).

9. See Londa Schiebinger and Claudia Swan, "Introduction," in Londa Schiebinger and Claudia Swan (eds.), *Colonial Botany: Science, Commerce, and Politics in the Early Modern World* (Philadelphia: University of Pennsylvania Press, 2007), pp. 7–13.

10. Londa Schiebinger, "Prospecting for Drugs: European Naturalists in the West Indies," ibid., pp. 105–117—here at p. 105.

11. One such illustration is offered by Patricia Fara in her *Sex, Botany and Empire: The Story of Carl Linnaeus and Joseph Banks* (London: Icon Books, 2017).

12. Georg Christian Oeder, quoted in Ib Friis, "G. C. Oeder's Conflict with Linnaeus and the Implementation of Taxonomic and Nomenclatural Ideas in the Monumental Flora Danica Project (1761–1883)," *Gardens' Bulletin Singapore* 71, suppl. 2 (2019): 53–85—here on p. 79.

13. Sigrun Skogly, *Beyond National Borders: States' Human Rights Obligations in International Cooperation* (Antwerp: Intersentia, 2006), p. 11.

14. Definition of bioprospecting, ScienceDirect, https://www.sciencedirect.com/topics/medicine-and-dentistry/bioprospecting. Accessed on 1August 18, 2021.

15. Staffan Müller-Wille, "Walnuts at Hudson Bay, Coral Reefs in Gotland: The Colonialism of Linnaean Botany," in Schiebinger and Swan (eds.), *Colonial Botany*, pp. 34–46—here at p. 46.

16. Schiebinger and Swan, "Introduction," p. 12.

17. Ibid., p. 8.

18. Georg Christian Oeder, quoted in Ib Friis, "G.C. Oeder's Conflict with Linnaeus and the Implementation of Taxonomic and Nomenclatural Ideas in the Monumental Flora Danica Project (1761–1883)," *Gardens' Bulletin Singapore* 71, suppl. 2 (2019): 53–85—here at p. 77.

19. None of the available sources mentions anything about why plants from the colonies outside Europe were not going to be a part of the *Flora Danica* project.

20. Oeder, quoted in Friis, "G.C. Oeder's Conflict with Linnaeus," p. 79.

21. Martin Vahl, *Eclogae Americanae I–IV* (1796–1807).

22. The title of the report was *Geschichte der Mission der evangelischen Brüder auf den caraibischen Inseln S. Thomas, S. Croix und S. Jan.* An English translation was published in 1987 entitled "A Caribbean Mission." "Oldendorp: The Man Behind an Impressive Publication About the Danish Colony," Danish National Archives, The Danish West-Indies: Sources of History, https://www.virgin-islands-history.org/en/history/fates/oldendorp-the-man-behind-an-impressive-publication-about-the-danish-colony/. Accessed on 1August 18, 2021.

23. See, e.g., Michael T. Bravo, "Missionary Gardens: Natural History and Global Expansion, 1720-1820," in Schiebinger and Swan, *Colonial Botany*, pp. 47–61.

24. See https://whc.unesco.org/en/list/1468/. Accessed on 1August 18, 2021.

25. Michael T. Bravo, "Missionary Gardens: Natural History and Global Expansion, 1720–1820," p. 48.

26. See "Moravian Church," *Encyclopedia Britannica*, https://www.britannica.com/topic/Moravian-church. Accessed on 1August 18, 2021.

27. In *This Month in Moravian History*, no. 27 (January 2008), the background for the work begun in Greenland and in Saint Thomas in the Danish West Indies is described in this way: "The Moravian mission in Greenland was begun only a few months after the Herrnhut community sent out missionaries to St. Thomas. The reason for choosing St. Thomas and Greenland as their first mission fields was the personal acquaintance Count Zinzendorf [the founder of Herrnhut as a Christian community] had made with people from these countries. When Zinzendorf attended the coronation of Christian VI and his wife Sophia Magdalena as King and Queen of Denmark in 1731, he met Anthony, a slave from St. Thomas, as well as two Greenlanders." Available at http://www.moravianchurcharchives.org/thismonth/08%20jan%20greenland.pdf. Accessed on 1August 18, 2021.

28. Henning Knudsen, *The Story Behind Flora Danica* (Copenhagen: Lindhardt and Ringhof, 2016), p. 105.

29. *This Month in Moravian History*.

30. Bravo, "Missionary Gardens," p. 52.

31. Knudsen, *The Story Behind Flora Danica*, p. 105.

32. See Felicity Jensz, "The Publication and Reception of David Cranz's 1767 *History of Greenland*," *The Library: The Transactions of the Bibliographical Society* 13, no. 4 (2012): 457–472; and Bravo, "Missionary Gardens," p. 56.

33. Another famous example is the aboriginal populations of Australia, who knew basically every plausible plant and animal to eat, everything about the weather, the land, the seasons, the dangers (and still do): the immigrants relied on this or ignored it at their peril.

34. Bravo, "Missionary Gardens," p. 61.

35. Knudsen, *The Story Behind Flora Danica*, p. 105. This and the next paragraph build on Knudsen's work.

36. Ibid.

37. Edward Daniel Clarke, quoted in Helge Kragh, "Mellem provinsialisme og internationalisme. Dansk naturvidenskab i oplysningstiden," in Marita Akjhøj Nielsen, (ed.) *Det fremmede som historisk drivkraft Danmark efter 1742. Et festskrift til Hendes Majestæt Dronning Margrethe II ved 70-års-fødselsdagen den 16. april 2010* (Copenhagen: Det Kongelige Danske Videnskabernes Selskab, 2010), pp. 26–34—here at p. 34. Available at https://botaniskforening.dk/wp-content/uploads/2015/12/botanik_dronningesymposium.pdf. Accessed on August 18, 2021.

38. Ibid., p. 26.

39. Ibid., p. 27.

40. Ib Friis, "Hvor dansk er den danske flora?" (How Danish is the Danish Flora?), in Marita Akjhøj Nielsen, (ed.) *Det fremmede som historisk drivkraft Danmark efter 1742. Et festskrift til Hendes Majestæt Dronning Margrethe II ved 70-års-fødselsdagen den 16. april 2010* (Copenhagen: Det Kongelige Danske Videnskabernes Selskab, 2010), pp. 13–25—here at p. 14. Available at https://botaniskforening.dk/wp-content/uploads/2015/12/botanik_dronningesymposium.pdf. Accessed on August 18, 2021.

41. Ibid.

42. Ibid.

43. Ibid., p. 17.

44. Invasive Alien Species, European Commission, https://ec.europa.eu/environment/nature/invasivealien/index_en.htm. Accessed on August 18, 2021.

45. Regulation (EU) No. 1143/2014 of the European Parliament and of the Council of 22 October 2014 on the prevention and management of the introduction and spread of invasive alien species, (1), Available at https://eur-lex.europa.eu/legal-content/EN/TXT/HTML/?uri=CELEX:32014R1143&from=EN. Accessed on August 18, 2021.

46. Friis, "Hvor dansk er den danske flora?," p. 21.

47. Ibid., p. 16.

48. Flora Nordica, https://www.nhbs.com/3/series/flora-nordica. Accessed on August 18, 2021. Presently without funding, the project has so far resulted in only a few published volumes.

49. The IUCN is a membership union composed of both government and civil society organizations whose aim is to conserve nature and to accelerate the transition to sustainable development. See https://www.iucn.org/about. Accessed on August 18, 2021.

50. Friis, "Hvor dansk er den danske flora?," p. 23.

51. Ibid. See also Knudsen, *The Story Behind Flora Danica*, p. 212.

52. Chris D. Thomas, *Inheritors of the Earth: How Nature Is Thriving in an Age of Extinction* (New York: Public Affairs, 2017).

53. Ibid., pp. 7–8.

54. See, e.g., Christoph Kueffer, Review of Chris D. Thomas, *Inheritors of the Earth: How Nature Is Thriving in an Age of Extinction, Basic and Applied Ecology* 35 (2019): 13–17.

55. International Covenant on Civil and Political Rights, Article 2.

56. European Convention on Human Rights, Article 1.

57. Mark Gibney, *International Human Rights Law: Returning to Universal Principles* (Lanham, Md.: Rowman and Littlefield, 2008), pp. 10–11. Gibney's argument that the protection of human rights has nothing to do with citizenship and does not stop at any territorial boundaries is an important one. It represents, he writes, "a fundamental misunderstanding of international human rights law itself," because "if human rights protection were something that individual states could (and would) do individually, there would be no need for any international convention" (p. 11).

58. Article 4 of the Convention of the Rights of the Child provides that "with regard to economic, social and cultural rights, States Parties shall undertake such measures to the maximum extent of their available resources and, where needed, within the framework of international cooperation." See also Skogly, *Beyond National Borders*, pp. 11–12.

59. CESCR General Comment No. 3: The Nature of States Parties' Obligations (Article 2, Paragraph 1 of the covenant), E/1991/23, 1991, para. 1 Available at https://www.refworld.org/pdfid/4538838e10.pdf. Accessed on August 18, 2021.

60. Ibid., para. 13.

61. Ibid., para. 14.

62. René Cassin quoted in Fons Coomans, "Application of the International Covenant on Economic, Social and Cultural Rights in the Framework of International Organisations," in A. von Bogdandy and R. Wolfrum, (eds.), *Max Planck Yearbook of United Nations Law*, Vol. 11, Max Planck Institute, 2007, pp. 359–390—here in note 11, pp. 362–63.

63. Ibid., p. 364.

64. Magdalena Sepúlveda Carmona, "The Obligations of 'International Assistance and Cooperation' Under the International Covenant on Economic, Social and Cultural Rights. A

Possible Entry Point to a Human Rights Based Approach to Millennium Development Goal 8," *International Journal of Human Rights* 13, no. 1 (2009): 86–109—here at p. 88.

65. CESCR General Comment No. 25 (2020) on Science and Economic, Social and Cultural Rights (Articles 15(1)(b), 15(2), 15(3), and 15(4) of the ICESCR), E/C.12/GC/25.

66. Ibid., para. 78.

67. Ibid., para. 79.

68. Ibid., para. 80.

69. Ibid., para. 81.

70. Ibid., para. 82.

71. Ibid.

72. Ibid., para. 84.

73. Ibid., para. 83.

74. SAR's Academic Freedom Monitoring Project, https://www.scholarsatrisk.org/academic-freedom-monitoring-project-index/. Accessed on August 18, 2021.

75. Ibid.

76. UNESCO Recommendation on Science and Scientific Researchers (2017), Article 31.

77. Ibid., Article 31.

78. Ibid., Article 11. This and the following few paragraphs are based on Sebastian Porsdam Mann, Maximilian M. Schmid, Peter Treit, and Helle Porsdam, "Scientific Freedom: A Constitutive Element of the Right to Science," in preparation.

79. Ibid., Article 11(a).

80. Ibid., Preamble.

81. Ibid., Article 18.

82. Ibid., Article 20.

83. Ibid., Article 14.

84. The 1974 Recommendation on the Status of Scientific Researchers is available at http://portal.unesco.org/en/ev.php-URL_ID=13131&URL_DO=DO_TOPIC&URL_SECTION=201.html. Accessed on 1August 18, 2021. The 2017 recommendation is a revision of the 1974 recommendation: "The General Conference of UNESCO decided at its 37th session in November 2013 (37C/Resolution40) that the Recommendation on the Status of Scientific Researchers (1974) should be revised to reflect the contemporary ethical and regulatory challenges relating to the governance of science and science-society relationship, taking account, inter alia, of the Declaration on Science and the Use of Scientific Knowledge adopted by the World Conference on Science on 1 July 1999 and the 2005 UNESCO Universal Declaration on Bioethics and Human Rights, in order to provide a powerful and relevant statement of science ethics as the basis for science policies that would favour the creation of an institutional order conducive to the realization of Article 27, paragraph 1 of the Universal Declaration of Human Rights." Summary, Revision of the Recommendation on the Status of Scientific Researchers (1974), https://unesdoc.unesco.org/ark:/48223/pf0000248910.

85. 2017 UNESCO Recommendation on Science and Scientific Researchers, Preamble: "Recognizing also: (a) the significant value of science as a common good."

86. United Nations Transforming Our World: The 2030 Agenda for Sustainable Development, A/RES/70/1 (2015). Available at https://sustainabledevelopment.un.org/content/documents/21252030%20Agenda%20for%20Sustainable%20Development%20web.pdf. Accessed on August 18, 2021.

87. UNESCO Recommendation on Science and Scientific Researchers (2017), Article 18.

88. General Comment No. 25, paras. 40–41.

89. Martin Fredriksson, "From Biopiracy to Bioprospecting: Negotiating the Limits of Propertization," in Martin Fredriksson and James Arvanitakis (eds.), *Property, Place and Piracy* (London: Routledge, 2017), pp. 174–186—here at 174.

90. See, e.g., World Intellectual Property Report 2017, "Intangible Capital in Value Chains," WIPO 2017. Available at https://www.wipo.int/edocs/pubdocs/en/wipo_pub_944_2017.pdf. Accessed on 1August 18, 2021.

91. Fredriksson, "From Biopiracy to Bioprospecting," p. 175.

92. Ibid., pp. 175–176.

93. For a more in-depth treatment of this topic, see my *The Transforming Power of Cultural Rights: A Promising Law and Humanities Approach* (Cambridge, U.K.: Cambridge University Press), Chapter 6.

94. United Nations Declaration on the Rights of Indigenous Peoples (2007), UN Document A/RES/61/295.

95. Ibid., Article 31.

96. 2017 UNESCO Recommendation, Article 16(b)(iii).

97. Mauro Barelli, "The United Nations Declaration on the Rights of Indigenous Peoples: A Human Rights Framework for Intellectual Property Rights," in M. Rimmer (ed.), *Indigenous Intellectual Property: A Handbook of Contemporary Research* (Cheltenham, U.K.: Edward Elgar 2017), pp. 47–63—here at p. 61.

98. See Introduction to the CBD, https://www.cbd.int/intro/. Accessed on 18 Auguat 2021.

99. Fredriksson, "From Biopiracy to Bioprospecting," p. 183.

100. See, e.g., Sheila Jasanoff (ed.), *States of Knowledge: The Co-production of Science and Social Order* (London: Routledge, 2004).

101. See James A. Secord, "Knowledge in Transit," *Isis* 95 (2004): 654–672; Sebastian Conrad, *What Is Global History?* (Princeton, N.J.: Princeton University Press, 2016).

102. I want to thank Jahnavi Phalkey for drawing my attention to this.

103. Thomas Efferth et al., "Biopiracy of Natural Products and Good Bioprospecting Practices," *Phytomedicine* 23 (2016): 166–173.

104. Ibid., p. 171.

105. Ibid., p. 172.

106. Ibid.

107. Ibid.

108. See Roberto Andorno, "The Right to Science and the Evolution of Scientific Integrity," in Helle Porsdam and Sebastian Porsdam Mann (eds.), *The Right to Science: Then and Now* (New York: Cambridge University Press, 2021), pp. 91–103.

109. Ibid., pp. 96–97.

110. *Proposals for Safeguarding Good Scientific Practice*, Deutsche Forschungsgemeinschaft, Bonn, 1998; Max-Planck-Society, *Rules of Good Scientific Practice & Rules of Procedure in Cases of Suspected Misconduct*. Max Planck Society, Munich, 2000 (revised in 2009). See also Andorno, pp. 9–10.

111. M. Nylenna, D. Andersen, G. Dahlquist, M. Sarvas, and A. Aakvaag. "Handling of Scientific Dishonesty in the Nordic Countries. National Committees on Scientific Dishonesty in the Nordic Countries," *Lancet*, 354, no. 9172 (1999): 57–61.

112. Ibid. See also D. Andersen, "From Case Management to Prevention of Scientific Dishonesty in Denmark," *Science and Engineering Ethics* 6, no. 1 (2000): 25–34.

113. R. Andorno, "The Right to Science and the Evolution of Scientific Integrity," p. 95.

114. Ibid., p. 96.

115. Naomi Oreskes, *Why Trust Science?* (Princeton, N.J.: Princeton University Press, 2019). See also Zoë Corbyn, "Naomi Oreskes: 'Discrediting Science Is a Political Strategy,'" *The Observer*, 3 November 2019 Available at https://www.theguardian.com/science/2019/nov /03/naomi-oreskes-interview-why-trust-science-climate-donald-trump-vaccine. Accessed on August 18, 2021.

116. In *The Story Behind Flora Danica*, Henning Knudsen calls the part, in which he describes and shows illustrations from *Flora Danica* of both Ramsons and Ground Elder, "the dynamic flora," pp. 155–237.

117. Ibid., p. 214. Finn T. Sørensen, "Skvalderkål" [ground elder], Meddelelser fra have-brugshistorisk Selskab, no. 24 (1994): 63–68.

118. List of plants under observation, Danish Ministry of Environment and Food, p. 51. Available at https://naturstyrelsen.dk/media/nst/66891/HandlingsplanForInvasiveArter.pdf. Accessed on August 18, 2021.

119. Amy McGuire, "Ramson and Friends," blog post, May 14, 2014, Nordic Food Lab, http://nordicfoodlab.org/blog/2014/5/ramson-and-friends. Accessed on August 18, 2021.

120. "Ny forskning: Nordisk-madbølgen udspringer i Grønland," *Videnskab.dk*, December 2, 2013, https://videnskab.dk/kort-nyt/ny-forskning-nordisk-madbolgen-udspringer-i -gronland. Accessed on August 18, 2021.

121. Ibid.

Chapter 6

1. *Transforming Our World: The 2030 Agenda for Sustainable Development*, A/RES/70/1, October 21, 2015, Paragraph 10. Available at https://www.un.org/ga/search/view_doc.asp ?symbol=A/RES/70/1&Lang=E. Accessed on August 18, 2021.

2. Ibid., Preamble.

3. "Science for Sustainable Development," policy brief by the Scientific Advisory Board of the Secretary General of the United Nations, October 5, 2016, SC/2016/UNSAB/ScDev. Available at https://www.iau-hesd.net/sites/default/files/documents/sciences-for-sd.pdf. Accessed on August 18, 2021.

4. "Science for Sustainable Development, 2. Guiding Principles: Science for Sustainable Development, Principle 1: Recognize Science as a Universal Public Good."

5. Derek Osborn, Amy Cutter, and Farooq Ullah, *Universal Sustainable Development Goals: Understanding the Transformational Challenge for Developed Countries.*, report of a study by Stakeholder Forum, May 2015. Available at https://sustainabledevelopment.un.org/content/documents /1684SF_-_SDG_Universality_Report_-_May_2015.pdf. Accessed on August 18, 2021.

6. See the UN publication, *COVID-10 and Human Rights: We Are All in This Together*, 2020, https://www.un.org/sites/un2.un.org/files/un_policy_brief_on_human_rights_and_covid_23 _april_2020.pdf. Accessed on August 18, 2021.

7. *The Road to Dignity by 2030: Ending Poverty, Transforming All Lives and Protecting the Planet*, synthesis report of the Secretary-General on the Post-2015 Agenda, December 2014, Paragraph 48, http://www.un.org/disabilities/documents/reports/SG_Synthesis_Report_Road _to_Dignity_by_2030.pdf. Accessed on August 18, 2021.

8. See Stjepan Oreskovic and Sebastian Porsdam Mann, "Science in the Times of SARS-CoV-2," in Helle Porsdam and Sebastian Porsdam Mann (eds,), *The Right to Science: Then and Now* (Cambridge, U.K.: Cambridge University Press, 2021), pp. 166–194.

9. Ibid. Oreskovic and Porsdam Mann mention as examples of altruistic behavior among states how a German university hospital admitted COVID-19 patients from France, how Albania sent a group of thirty doctors and nurses to neighboring Italy, and how France responded forcefully to the need for more blood donations.

10. See, e.g., Onora O'Neill, "Justice Without Ethics: A Twentieth Century Innovation?," in John Tasioulas (ed.), *A Cambridge Companion to the Philosophy of Law* (Cambridge, U.K.: Cambridge University Press, 2020), pp. 135–151.

11. *A Universal Declaration on Human Responsibilities*, InterAction Council, 1997. Available at https://www.interactioncouncil.org/publications/universal-declaration-human -responsibilities. Accessed on August 18, 2021.

12. O'Neill, "Justice Without Ethics: A Twentieth Century Innovation?"

13. A variety of definitions are used for "science diplomacy." I use the one offered by SciTech DiploHub: http://www.scitechdiplohub.org/what-is-science-diplomacy/. Accessed on August 18, 2021.

14. "Jennifer Doudna on How Covid-19 Is Spurring Science to Accelerate," *The Economist*, June 5, 2020, https://www.economist.com/by-invitation/2020/06/05/jennifer-doudna-on-how -covid-19-is-spurring-science-to-accelerate. Accessed on August 18, 2021.

15. See Oreskovic and Porsdam Mann, "Science in the Times of SARS-CoV-2."

16. A. G. "The Greenland Flora," *Botanical Gazette* 7, no. 3 (1882), p. 27.

17. Personal communication from James Ellis, managing editor of the *International Journal of Plant Sciences*. I thank Ellis for taking the time to engage with me on this issue. On the life and importance of Asa Gray see Melissa Petruzzello, "Asa Gray: The Father of American Botany," *Encyclopedia Britannica*, https://www.britannica.com/story/asa-gray-the-father-of-american -botany. Accessed on August 18, 2021.

18. "Library at the Botanical Garden," in *Public Libraries in the United States of America*, p. 88. Available at https://books.google.dk/books?id=RWMsAAAAYAAJ&pg=PA88&lpg = PA88 & dq = Asa + Gray + on + Flora + Danica & source = bl & ots = xa2Cpf-Psb & sig = ACfU3U3u7uGyZTTNnHfGIj0hR4OyT_z3Jw & hl = da & sa = X & ved = 2ahUKEwjA1f LeqJPqAhXVxcQBHVTsCx0Q6AEwCnoECAsQAQ#v=onepage&q=Asa%20Gray%20on %20Flora%20Danica&f=false. Accessed on August 18, 2021.

19. Asa Gray quoted in "Asa Gray," *Proceedings of the American Academy of Arts and Sciences* 23, no. 2 (May 1887–May 1888), pp. 321–343—here at 337.

20. Personal communication from Donald H. Pfister, curator of the Farlow Library and Herbarium of Cryptoganic Botany and Asa Gray Professor of Systematic Botany at Harvard University. I thank Professor Pfister for drawing my attention to this text and this possible explanation for Gray's remarks.

21. Peter Wagner, "The Royal Work *Flora Danica*," in *Flora Danica og det Danske hof (Flora Danica and the Royal Copenhagen Porcelain Manufactory)*. Udstilling af porcelæns-, guld- og sølvservice (Copenhagen: Christiansborg Castle, 1990), pp. 15–44—here at p. 32.

22. Henning Knudsen, *The Story Behind Flora Danica* (Copenhagen: Lindhardt and Ringhof, 2016), p. 26. Because of Müller's work on fungi, algae, and zoology, Knudsen considers him the "greatest [Danish] naturalist of all times."

23. Ibid., p. 28.

24. Martin Vahl, *Eclogae Americanae I–IV* (1796–1807).

25. For more on the Arabian exhibition, also known as the Carsten Niebuhr Expedition, see, e.g., Paul G. Chamberlain, "Carsten Niebuhr and the Danish Expedition to Arabia," https://

www.aramcoworld.com/Articles/January-2018/Carsten-Niebuhr-and-the-Danish-Expedition-to-Arabi. Accessed on August 18, 2021.

Containing more than 1,400 new species, "this collection is one of the most valuable ones from a scientific point of view at the Natural History Museum of Denmark," writes Henning Knudsen (*The Story Behind Flora Danica*, p. 28).

26. Knudsen, *The Story Behind Flora Danica*, p. 27.

27. Ibid., p. 28.

28. Danish Natural History Museum, "Martin Vahl's manuscript" (in Danish), https://samlinger.snm.ku.dk/toer-og-vaadsamlinger/botanik/general-herbarium/martin-vahls-1749-1804-manuscript. Accessed on August 18, 2021.

29. Wagner, "The Royal Work *Flora Danica*," p. 32.

30. Ibid., p. 34.

31. Jean Anker, "From the Early History of the Flora Danica," *Libri* 1 (1951): 334–350—here at p. 350.

32. Wagner, "The Royal Work *Flora Danica*," p. 34.

33. Liebmann even married Hornemann's youngest daughter. Hornemann was also the uncle of Johan Lange, the last *Flora Danica* editor.

34. Knudsen, *The Story Behind Flora Danica*, p. 35.

35. Ibid., p. 36.

36. Ibid., p. 35.

37. Ibid., p. 38; Wagner, "The Royal Work *Flora Danica*," p. 36.

38. Proceedings (Yearbook) of the Royal Danish Academy of Sciences and Letters, 1874, archival no. 998, p. 35. My own translation from the Danish.

39. Proceedings (Yearbook) of the Royal Danish Academy of Sciences and Letters, 1880, archival no. 497, p. 111. My own translation from the Danish.

40. Knudsen, *The Story Behind Flora Danica*, p. 592. Knudsen lists twenty-three of the most important of the these "florists" in his book.

41. Müller had also made inquiries as early as 1776. He had found only two copies of *Flora Danica* "in the hands of people who were interested in botany. The remainder must have got into the wrong hands or not been delivered at all" (Wagner, "The Royal Work *Flora Danica*," p. 34).

42. Revised plan for the work, approved by Christian VIII in 1842, quoted in Wagner, "The Royal Work *Flora Danica*," p. 34.

43. In 1808, German botanist and physician Kurt Sprengel wrote, "All of our science is illuminated by a remarkable light from Denmark now that a botanical treasure chest is being issued containing an exact examination of all the plants of that realm." Quoted in Knudsen, *The Story Behind Flora Danica*, p. 38.

44. Wagner, "The Royal Work *Flora Danica*," p. 38.

45. Knudsen, *The Story Behind Flora Danica*, p. 588.

46. Ib Friis, "G. C. Oeder's Conflict with Linnaeus and the Implementation of Taxonomic and Nomenclatural Ideas in the Monumental Flora Danica Project (1761–1883)," *Gardens' Bulletin Singapore* 71, suppl. 2 (2019): 53–85. Appendix 1. Oeder's invitation to subscribe to a new work called *Flora Danica* (1761), p. 79.

47. Anker, "From the Early History of the Flora Danica," p. 350.

48. Knudsen, *The Story Behind Flora Danica*, p. 39.

49. Declaration of Independence, 4 July 1776, https://www.archives.gov/founding-docs/declaration-transcript. Accessed on August 18, 2021.

50. Lorraine Daston and Michael Stolleis (eds.), *Natural Law and Laws of Nature in Early Modern Europe: Jurisprudence, Theology, Moral and Natural Philosophy* (London: Routledge, 2008), p. 12.

51. Burns Weston, "Human Rights in Henry J. Steiner and Philip Alston (eds), *International Human Rights in Context: Law, Politics, Morals* (New York: Oxford University Press, 2nd edition, 2000), pp. 324–326—here at p. 324.

52. In *The Last Utopia: Human Rights in History* (Cambridge, Mass.: Harvard University Press, 2012), Samuel Moyn argues that human rights did not become a part of genuine social movements until the 1970s. Other human rights historians have maintained that the true human rights moment was World War II and the founding of the UN in 1945.

53. Jeremy Bentham, "Rights, Representation, and Reform: Nonsense upon Stilts and Other Writings on the French Revolution," in P. Schofield, C. Pease-Watkin, and C. Blamires (eds.), *The Collected Works of Jeremy Bentham* (New York: Oxford University Press, 2002), pp. 317–401.

54. Weston, "Human Rights," p. 326.

55. Malcolm Langford, "Critiques of Human Rights," *Annual Review of Law and Social Science* 14 (2018): 69–89.

56. Ibid., pp. 72–75.

57. Ibid., pp. 75–79.

58. Ibid., pp. 79–82.

59. Kathryn Sikkink, *The Hidden Face of Rights: Toward a Politics of Responsibilities* (New Haven, Conn.: Yale University Press, 2020), pp. 16, 52.

60. Ibid., p. 25.

61. Onora O'Neill, "Justice Without Ethics: A Twentieth Century Innovation?," in John Tasioulas (ed.), *The Cambridge Companion to the Philosophy of Law* (Cambridge, U.K.: Cambridge University Press, 2020), pp. 135–151.

62. Ibid., p. 151.

63. Ibid., p. 135.

64. See list of council members and supporters at the end of the "Universal Declaration on Human Responsibilities," 1997.

65. Ibid., InterAction Council, "Introductory comment."

66. Chapter 2, "Duties," concerns the duties that "every individual shall have . . . towards his family and society, the State and other legally recognized communities and the international community." It also specifies that individual rights and freedoms must be balanced against those of others, with "due regard to . . . collective security, morality and common interest." African Charter on Human and Peoples' Rights, 1981. Available at http://www.hrcr.org/docs/Banjul/afrhr5.html. Accessed on August 18, 2021.

67. See Sikkink, *The Hidden Face of Rights*, pp. 31–32.

68. Langford, "Critiques of Human Rights," p. 83.

69. OHCHR position paper, "Transforming Our World: Human Rights in the 2030 Agenda for Sustainable Development," p. 2. Available at https://www.ohchr.org/Documents/Issues/MDGs/Post2015/HRAndPost2015.pdf. Accessed on August 18, 2021.

70. CESCR statement, "The Pledge to Leave No One Behind: The International Covenant on Economic, Social and Cultural Rights and the 2030 Agenda for Sustainable Development," E/C.12/2019/1, 9 April 2019, paragraph 1. Available at https://sustainabledevelopment.un.org/content/documents/21780E_C.12_2019_1_edited.pdf. Accessed on August 18, 2021.

71. Ibid., para. 5.

72. Ibid., paras. 16, 17.

73. Irina Bokova, "Foreword," *UNESCO Science Report: Towards 2030*, 2015, p. xx. Available at https://unesdoc.unesco.org/ark:/48223/pf0000235406/PDF/235406eng.pdf.multi. Accessed on August 18, 2021.

74. Article 22 UDHR reads, "Everyone, as a member of society, has the right to social security and is entitled to realization, through national effort and international co-operation and in accordance with the organization and resources of each State, of the economic, social and cultural rights indispensable for his dignity and the free development of his personality."

75. Johannes Morsink, *The Universal Declaration of Human Rights: Origins, Drafting and Intent* (Philadelphia: University of Pennsylvania Press, 1999), p. 212.

76. See Laurence R. Helfer and Graeme W. Austin, *Human Rights and Intellectual Property: Mapping the Global Interface* (Cambridge, U.K.: Cambridge University Press, 2011), p. 144.

77. "Science for Sustainable Development," Principle 1: Recognize science as a universal public good.

78. "Applied sciences and basic sciences are equally important for sustainable development and should not be played out against each other; they are two sides of the same coin." "Science for Sustainable Development," 2. Guiding Principles: Science for Sustainable Development, Principle 2: Acknowledge Basic Science as a Principal Requirement for Innovation.

79. "Science for Sustainable Development," Principle 1: Recognize science as a universal public good.

80. See Flurina Schneider, Andreas Kläy, Anne B. Zimmermann, Tobias Buser, Micah Ingalls, and Peter Messerli, "How Can Science Support the 2030 Agenda for Sustainable Development? Four Tasks to Tackle the Normative Dimension of Sustainability," *Sustainability Science* 14 (2019): 1593–1604—here at p. 1594. Available at https://doi.org/10.1007/s11625-019-00675-y).

81. https://en.unesco.org/news/sesame-facility-slated-boost-science-and-cooperation-middle-east. Accessed on August 18, 2021.

82. https://www.sciencediplomacy.org/. Accessed on July 24, 2020.

83. https://royalsociety.org/topics-policy/publications/2010/new-frontiers-science-diplomacy/. Accessed on August 18, 2021.

84. Carlos Moedas, "Science Diplomacy in the European Union," *Science & Diplomacy* 5, no. 1 (March 2016), http://www.sciencediplomacy.org/perspective/2016/science-diplomacy-in-european-union. Accessed on August 18, 2021.

85. German Federal Foreign Office, "Science Diplomacy: A New Strategy in Research and Academic Relations Policy," December 2020. Available at https://www.auswaertiges-amt.de/blob/2436494/2b868e9f63a4f5ffe703faba680a61c0/201203-science-diplomacy-strategiepapier-data.pdf. Accessed on August 18, 2021.

86. Ibid., p. 2.

87. Ibid., p. 1.

88. Derek Massarella, "Philip Henry Zollman, The Royal Society's First Assistant Secretary for Foreign Correspondence," *Notes and Records* 46, no. 2 (1992): 219–234, here at p. 219. Available at https://royalsocietypublishing.org/doi/pdf/10.1098/rsnr.1992.0023. Accessed on August 18, 2021.

89. Ibid. See also the Royal Society, *New Frontiers in Science Diplomacy: Navigating the Changing Balance of Power*, Policy document, January 2012. Available at https://royalsociety.org/-/media/Royal_Society_Content/policy/publications/2010/4294969468.pdf. Accessed on August 18, 2021.

90. Roberto Andorno, "The role of UNESCO in Promoting Universal Human Rights: From 1948 to 2005," in C. Caporale and I. Pavone (eds.), *International Biolaw and Shared Ethical Principles. The Universal Declaration on Bioethics and Human Rights* (Oxford: Routledge, 2018), pp. 7–21.

91. "Russell-Einstein Manifesto," London, 9 July 1955. Available at https://www
.atomicheritage.org/key-documents/russell-einstein-manifesto. Accessed on August 18, 2021.

92. Pugwash, https://pugwash.org/about-pugwash/. Accessed on August 18, 2021.

93. Bohr, quoted in Finn Aaserud, "Niels Bohr's Mission for an 'Open World,'" in M.
Kokowski (ed.), *The Global and the Local: The History of Science and the Cultural Integration of
Europe. Proceedings of the 2nd ICESHS*, 2006, pp. 706–709—here at p. 708.

94. Finn Aaserud, "Niels Bohr's Diplomatic Mission During and After World War Two,"
Berichte zur Wissenschaftsgeschichte 43, no. 4 (2020): 493–520.

95. Niels Bohr, "Open Letter to the United Nations," 1950. Available at http://www
.atomicarchive.com/Docs/Deterrence/BohrUN.shtml. Accessed on August 18, 2021.

96. Aaserud, "Statesmen and Diplomats Encounter Niels Bohr."

97. Marc Busch and David Logsdon, quoted in Patty Nieberg, "The True Source of US-China
Trade Tension Is Over Technology," *Medill News Service*, 4 February 2019. Available at https://
dc.medill.northwestern.edu/blog/2019/02/04/the-true-source-of-u-s-china-trade-tension
-is-over-technology/#sthash.4iZqhzbU.dpbs. Accessed on 14 August 2021.

98. Peter Gluckman and Vaughan Turekian. "Rebooting Science Diplomacy in the Context of COVID-19," *Issues in Science and Technology* (June 17, 2020). Available at https://issues
.org/rebooting-science-diplomacy-in-the-context-of-covid-19-lessons-from-the-cold-war/.
Accessed on August 18, 2021.

99. As one of his first acts in office, in January 2021 President Joe Biden signed an executive
order reversing his predecessor's decision to leave the WHO.

100. Nieburg, "The True Source of US-China Trade Tension Is Technology."

101. Ibid.

102. Gluckman and Turekian, "Rebooting Science Diplomacy in the Context of COVID-19."

103. "China's Arctic Policy," The State Council Information Office of the People's Republic
of China, January 2018. Available at http://english.www.gov.cn/archive/white_paper/2018/01
/26/content_281476026660336.htm. Accessed on August 18, 2021.

104. The Arctic Council, https://arctic-council.org/en/about/. Accessed on August 18, 2021.

105. "Joint Communique of the Governments of the Arctic Countries on the Establishment of the Arctic Council," Ottawa, Canada, 19 September 1996. Available at https://oaarchive
.arctic-council.org/bitstream/handle/11374/85/EDOCS-1752-v2-ACMMCA00_Ottawa_1996
_Founding_Declaration.PDF?sequence=5&isAllowed=y. Accessed on August 18, 2021.

106. Founding Articles for an International Arctic Science Committee (IASC), August
1990. Available from https://iasc.info/images/about/iasc-founding-articles.pdf. Accessed on
August 18, 2021.

107. IASC, https://iasc.info/iasc/about-iasc.

108. "China's Arctic Policy," II. China and the Arctic.

109. Atle Staalesen, "A New Global Order Is Coming to the Arctic: Strong Voices Say It Must
Be Met by an Overhaul in Regional Governance," *Barents Observer*, 29 January 2020. Available
at https://thebarentsobserver.com/en/arctic/2020/01/new-global-order-coming-arctic-strong
-voices-say-it-must-be-met-overhaul-regional. Accessed on August 18, 2021.

110. See, e.g., Jon Rahbek-Clemmensen, "When Do Ideas of an Arctic Treaty Become
Prominent in Arctic Governance Debates?" *Arctic* 72, no. 2 (2019): 116–130. Available at www
.jstor.org/stable/26739923. Accessed on August 18, 2021.

111. The Antarctic Treaty, https://www.ats.aq/e/antarctictreaty.html. Accessed on
August 18, 2021.

112. Ibid.

113. Marc Lanteigne, "So You Want to Write an Arctic Treaty?," *Over the Circle: Arctic News and Analysis*, 10 February 2020, https://overthecircle.com/2020/02/10/so-you-want-to-write-an -arctic-treaty/. Accessed on August 18, 2021.

114. John B. Bellinger, "Treaty on Ice," *New York Times*, June 23, 2008. Available at https:// www.nytimes.com/2008/06/23/opinion/23bellinger.html. Accessed on August 18, 2021.

115. Lanteigne, "So You Want to Write an Arctic Treaty?"

116. Neil Shea, "Scenes from the New Cold War Unfolding at the Top of the World," *National Geographic*, May 8, 2019. Available at https://www.nationalgeographic.com/environment/2018 /10/new-cold-war-brews-as-arctic-ice-melts/. Accessed on August 18, 2021.

117. Johan Lange, "Fortale" (Introduction), *Oversigt over Grønlands flora* (Survey of Green-land's Flora) (Copenhagen: Hans Reitzel, 2nd edition, 1890). My own translation from the Dan-ish. Available at https://www.biodiversitylibrary.org/item/263465#page/15/mode/1up. Accessed on July 29, 2020.

Conclusion

1. In Danish, *plattedamerne*.

2. The process is explained here: http://floradanica.royalcopenhagen.com/the-craft/. Accessed on August 18, 2021.

3. Lise Togeby, "Feminist Attitudes in Times of Depoliticization of Women's Issues," *Euro-pean Journal of Political Research* 27 (1995): 41–68.

4. UNESCO, Priority Gender Equality: https://en.unesco.org/genderequality. Accessed on August 18, 2021.

5. UN Office of the High Commissioner, "About the Mandate of the Special Rappor-teur in the Field of Cultural Rights," https://www.ohchr.org/EN/Issues/CulturalRights/Pages /MandateInfo.aspx. Accessed on August 18, 2021.

6. Ibid.

7. Report of the special rapporteur in the field of cultural rights, A/HRC/20/26, 2012. I also touch on Appadurai's politics of hope in Chapter 5 on the right to science in my *The Trans-forming Power of Cultural Rights: A Promising Law and Humanities Approach* (Cambridge, U.K.: Cambridge University Press, 2019).

8. Arjun Appadurai, *The Future as Cultural Artefact: Essays on the Global Condition* (Lon-don, Verso, 2013), p. 289.

9. Arjun Appadurai, "The Right to Research," *Globalisation, Societies and Education* 4, no. 2 (2006): 176–77.

10. See Sebastian Porsdam Mann, Helle Porsdam, and Yvonne Donders, "'Sleeping Beauty': The Right to Science as a Global Ethical Discourse," *Human Rights Quarterly* 42 (2020): 332– 335—here at p. 353.

11. "Charter of the Book," *UNESCO Bulletin for Libraries* 26, no. 5 (September–October 1972): 238–240.

12. Ibid., p. 238.

13. Joanna Sikoraa, M. D. R. Evans, and Jonathan Kelley, "Scholarly Culture: How Books in Adolescence Enhance Adult Literacy, Numeracy and Technology Skills in 31 Societies," *Social Science Research* 77 (2019): 1–15.

14. Ibid., p. 2.

15. Ibid., pp. 1–2.

16. Ibid., p. 3.

17. Ibid., p. 14.

18. Ibid., p. 8.

19. Ibid., p. 14.

20. Lea Shaver, *Ending Book Hunger: Access to Print Across Barriers of Class and Culture* (New Haven, Conn.: Yale University Press, 2019), p. 2.

21. Ibid., p. 17.

22. Shaheed calls Shaver a "catalyst" for her understanding of the right to science. Farida Shaheed, "The United Nations Cultural Rights Mandate: Reflections on the Significance and Challenges," in Lucky Belder and Helle Porsdam (eds.), *Negotiating Cultural Rights: Issues at Stake, Challenges and Recommendations* (London: Edward Elgar, 2017), pp. 21–36—here at p. 32.

23. Shaver, *Ending Book Hunger*, p. 9.

24. Ibid., p. 171.

25. Ibid., p. 172.

26. The rapid development of artificial intelligence–driven, automated translation systems familiar from Google Translate has considerable promise to solve this problem, especially as these models easily adapt to any combination of input and output languages. It will be interesting to see if copyrights on translations—in addition to the original text—would apply in this case.

27. Shaver, *Ending Book Hunger*, pp. 139, 183–184.

28. Ibid., pp. 183–184.

29. Kamaljit S. Bawa, Eben Goodale, Wambura Mtemi, You-Fang Chen, Ranjit Barthakur, Uromi Manage Goodale, Jianguo Liu, Aiwu Jiang, Christos Mammides, Madhava Meegaskumbura, Maharaj K. Pandit, and Kun-Fang Cao, "China and India: Toward a Sustainable World," *Science* 369, no. 6503 (2020): 515.

INDEX

AAAS. *See* American Association for the Advancement of Science
absolutism, abolition of, 128
absolutist king, men working for, 59
academia, 117–118, 123; global, 8; museums and, 71
Academic Freedom Monitoring Project, 111, 113
acceptability, 65
access: democratic, 93, 132; dialogue concerning, 36; facilitating for all, 68; furthering, 65; lack of, 127; less, 150; no, 83; open, 45, 64, 135–136; to and activate right to science, 6; to and distribution and dissemination of science, 64; to and share scientific knowledge, 2; to benefits, 23, 45, 93; to books, 149; to corporate affiliations, 36; to cultural resources, 148; to data, 36; to financial interests, 36; to intellectual property, 36; to knowledge, 3, 19, 31, 36, 45, 64, 112, 123; to medical advice, 61; to reading materials, 148; to research, 146; to science (and culture), 146; to technology, 32, 45, 134; to the internet, 64; policies on, 70; public, 45, 64; women's, 43. *See also* participation
accessibility, CESCR, 65
activity: creative, 22, 66, 78; hands-on, 23; research, 95; scientific, 3, 32, 73
actors: compete with, 57; foreign commercial, 114; fundamentalist, 48; governmental, 111; nonstate, 45, 66; political, 90
affiliations, corporate, 36
agriculture, 1, 5, 60, 79; Chinese economic interest in, 141; domestic policy objective, 19; investment in knowledge on, 59; modernization of, 17; utilizing plants for the support of, 17
Alechinsky, Pierre, 41
d'Alembert, Jean, 12

Alexandra, Princess, 40
Algiers, 103
Allkin, Bob, 84
almanacs, 63
alternative facts, 28
American Association for the Advancement of Science, 24, 136; Science and Human Rights Coalition, 7, 26; statement, 51
American Declaration of Independence, 131
American Declaration of the Rights and Duties of Man, drafting and adoption of, 35
Anasta, Paul, 49
Andersen, Hans Christian, 53, 97
Andorno, Roberto, 119
Anker, Jean, 14
Antarctic Treaty, 141
Anthropocene, 85, 101, 107; evolution, 107–108
antibiotics, 77
Appadurai, Arjun, 146–147
appropriation: illegitimate use and, 114; protected from misappropriation, 84
Arabian Peninsula, 126
Arctic Council, 140, 143
Arctic Treaty, 141
Arnold, Ken, 72
artificial intelligence, 110
artist, scientific, 38
Askov Folk High School, 68
Atomic Energy Commission, 51
atomic weapons, development of, 78
austerity measures, 66, 94
Australian Citizen Science Association, 73
authentication of herbs and plants, 83–84
authorship, 70; coauthorship, 95, 140; disagreements about, 119; scientific, 95
availability, CESCR, 65
awareness: ethical, 51; public, 69; raising, 111

CPSIA information can be obtained
at www.ICGtesting.com
Printed in the USA
JSHW061430120722
27850JS00003B/4